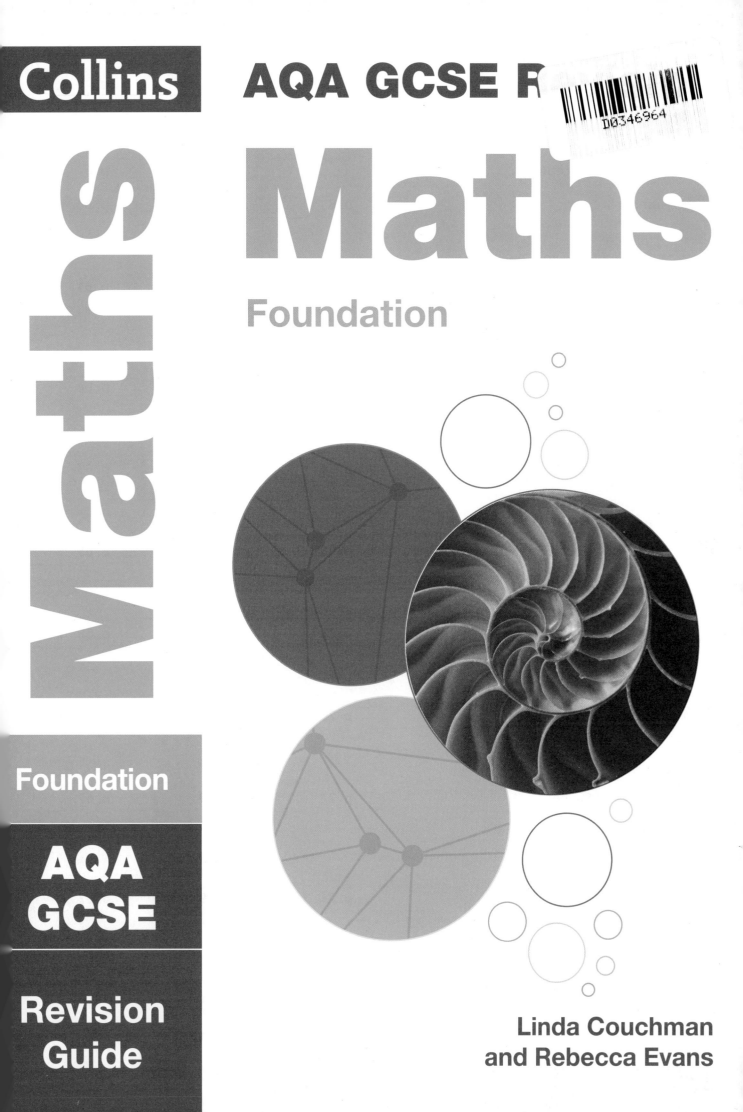

Collins

AQA GCSE R...

Maths

Foundation

Maths

Foundation

AQA
GCSE

Revision
Guide

Linda Couchman
and Rebecca Evans

Contents

N Number   **A** Algebra   **G** Geometry and Measures

Contents

Contents

N Number   **A** Algebra   **G** Geometry and Measures

Contents

		Revise	Practise	Review

Review Questions

Recap of KS3 Key Concepts

1 Write down two million and five as a number. [1]

2 Work out 15% of 300kg. [1]

3 Write 60 as the product of prime factors. [1]

4 Find **a)** the HCF and **b)** the LCM of 36 and 90. [2]

5 $6 + 4 \div 2 =$ _____ [1]

6 $(-2)^3 =$ _____ [1]

7 $(2^3)^2 =$ _____ [1]

8 Simplify: $5 + 2(m - 1)$ [1]

9 Simplify: $y^2 + y^2 + y^2$ [1]

10 **a)** Write the next two terms in the number sequence: 5, 8, 11, 14, __, __ [1]

b) Is 140 a term in the sequence? Explain your answer. [2]

11 $0.3 \times 0.2 =$ _____ [1]

12 A regular polygon has an interior angle of 140°. How many sides does the polygon have? [1]

13 Work out $4.152 \div 1.2$ [1]

14 If $A = (2, 3)$ and $B = (7, 11)$, work out the coordinates of the midpoint of the line AB. [2]

15 $3 \times 3 \times 3 =$ _____ [1]

16 Lorna left her home at 11.00am and went for a 20km run. She arrived back home at 1.00pm.

Work out Lorna's average running speed. [1]

17 The cross-section of a prism has an area of 25cm². The length of the prism is 5cm.

Work out the volume of the prism. [1]

18 The probability of it raining is 0.6

Work out the probability of it **not** raining. [1]

19 Write $\frac{7}{8}$ as a decimal. [1]

20 Write down 0.78 as a fraction in its simplest form. [2]

21 Find a fraction that lies between $\frac{2}{3}$ and $\frac{3}{4}$. [2]

22 Pat threw three darts. The lowest score was 10. The range was 10. The mean was 15.

Work out the score for each dart. [3]

23 Write down the number that is one less than one million. [1]

24 Share £56 in the ratio 3 : 5 [2]

25 Put the following numbers in order, smallest first: 0.4, 0.04, 0.39, 0.394 [1]

26 Calculate a value for m if:

a) $m - 5 = -1$ [1]

b) $m \div 4 = 12$ [1]

c) $2m + 13 = 16$ [1]

27 A square has a perimeter of 24cm. Work out its area. [1]

28 Work out $1616 \div 4$. [1]

29 Temi spends £7.25. She pays with a £20 note.

How much change does she receive? [1]

30 Simplify $10a - 6y + 3a - 4y$ [1]

31 Increase £60 by 35%. [1]

32 A circle has a radius of 10cm. Taking π as 3.14, work out the area of the circle. [2]

33 How many cubes of side length 2cm will fit inside a hollow cube of side length 4cm? [3]

34 Write down the square root of 225. [1]

35 $\frac{3}{4}$ of a number is 63. Write down the number. [2]

36 Work out the value of m if $2 - m = 5$. Circle the correct option:

7   3   −3   −7 [1]

37 Give the coordinates of the point where the graph of $x + y = 5$ crosses the x-axis. [2]

38 The two shorter sides of a right-angled triangle are 5cm and 12cm.

Work out the length of the third side. [1]

39 If $a = 3$ and $b = 4$, work out the value of $2a^2 + 3b$ [2]

40 Find the median of 1, 4, 6, 13, 21, 10, 3 and 7. [1]

Total Marks _____ / 57

Number 1

You must be able to:

- Use the four operations: addition, subtraction, multiplication and division
- Use BIDMAS
- Carry out calculations using a calculator.

The Four Operations

- You must be able to identify which operations are needed to answer a question.
- **Addition** problems might use words like: *sum*, *total*, *increase* and *plus*.
- **Subtraction** problems might use words like: *minus*, *decrease*, *take away* and *difference*.
- **Multiplication** problems might use words like: *multiply*, *times* and *product*.
- **Division** problems might use words like: *divide* and *share*.

> **Key Point**
>
> You may have to break some questions down and use a different operation at each stage.

Over five years, the population of a town increased by 3246 people.

If there were 15 628 people originally, how many were there at the end of the five-year period?

$$\begin{array}{r} 15628 \\ 3246 \, + \\ \hline 18874 \\ {\scriptstyle 1} \end{array}$$ 18874 people

A theatre can seat 600 people. On one night, 476 people watched a play.

How many seats were **not** filled?

$$\begin{array}{r} {\scriptstyle 5\;\;9\;\;1} \\ 6\cancel{0}\cancel{0} \\ 476 \, - \\ \hline 124 \end{array}$$ 124 seats

← Borrow across the top.

Cauliflowers are delivered to a supermarket in 48-kilogram crates. On Friday, 26 crates of cauliflowers were delivered.

What was the total weight of the delivery on Friday?

×	40	8
20	800	160
6	240	48

→

$$\begin{array}{r} 800 \\ 160 \\ 240 \\ 48 \, + \\ \hline 1248 \\ {\scriptstyle 1} \end{array}$$ 1248kg

← Multiply the tens and units and then add the products.

In an aquarium, a large tank can hold nine exotic fish.

How many tanks would be needed for 936 exotic fish?

$$\frac{104}{9\overline{)93^36}}$$ 104 tanks

9 goes into 9 once.
9 goes into 3 zero times, with 3 left over.
9 goes into 36 four times.

BIDMAS

- **BIDMAS** gives the order in which operations should be carried out:
    - **B**rackets (carry out the calculation in brackets first)
    - **I**ndices (roots and powers)
    - **D**ivision
    - **M**ultiplication
    - **A**ddition
    - **S**ubtraction.

Work out $6 + 4 \times 3$ $6 + 12 = 18$

Multiplication must be carried out before addition.

Using a Calculator

Use your calculator to work out $(18 + 37) \times 12$

ON (1 8 + 3 7) × 1 2 = 660

Press the calculator keys in this order.

Use your calculator to convert $\frac{3}{8}$ to a decimal.

ON 3 ÷ 8 = 0.375

Press S⇔D to switch between a fraction and a decimal answer.

Use your calculator to work out:

a) $\sqrt{1369}$

ON √■ 1 3 6 9 = 37

b) $\sqrt[3]{4096}$

ON SHIFT ³√■ 4 0 9 6 = 16

c) 6^4

ON 6 $x^■$ 4 = 1296

Key Point

Make sure you have practised using your calculator before the exams and know where all the main function keys are.

Quick Test

1. There are 12 greetings cards in a box. Raj needs to send 158 cards. How many boxes does he need to buy?
2. Mia says $2 + (3 + 4) + 3 \times 4 = 48$
   Molly says $2 + (3 + 4) + 3 \times 4 = 21$
   Who is correct? You must explain your answer clearly.

Key Words

addition
subtraction
multiplication
division
BIDMAS

Number 2

You must be able to:

- Order and compare positive and negative numbers
- Carry out calculations using positive and negative integers and decimals
- Understand and use standard form.

Ordering Numbers

- The value of each digit in a number depends on its position within that number. This is its **place value**.

Put the following numbers in order, from smallest to largest:
3400   34.03   340 000   34 030   340.3

Hundred Thousands	Ten Thousands	Thousands	Hundreds	Tens	Units		Tenths	Hundredths
		3	4	0	0			
				3	4	.	0	3
3	4	0	0	0	0			
	3	4	0	3	0			
			3	4	0	.	3	

In order, smallest to largest: 34.03, 340.3, 3400, 34 030, 340 000

- A number line can help to visualise questions.

Negative Numbers   **Positive Numbers**

−10 −9 −8 −7 −6 −5 −4 −3 −2 −1 0 1 2 3 4 5 6 7 8 9 10

Key Point

On a number line, positive numbers (+) are to the right of zero and negative numbers (−) are to the left of zero.

Numbers on the left are smaller than numbers on the right.

A negative number is smaller than zero.

Put the following numbers in order, smallest to largest:

−10.2   −0.3   −$\frac{1}{5}$ −6.4

−10.2   −0.3   −0.2   −6.4   ← Change all the numbers into decimals.

−10.2   −6.4   −0.3   −0.2   ← Then rearrange by size.

−10.2   −6.4   −0.3   −$\frac{1}{5}$ ← Use the original numbers in your final answer.

Calculating with Negative Numbers

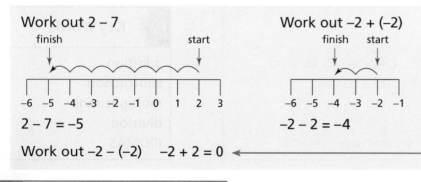

Work out 2 − 7

finish   start

−6 −5 −4 −3 −2 −1 0 1 2 3

2 − 7 = −5

Work out −2 + (−2)

finish   start

−6 −5 −4 −3 −2 −1

−2 − 2 = −4

Work out −2 − (−2)   −2 + 2 = 0

Two minus signs together become a plus.

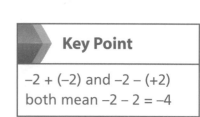

Key Point

−2 + (−2) and −2 − (+2) both mean −2 − 2 = −4

- When multiplying or dividing with two signs that are **different**, the answer is **negative**.
- When multiplying or dividing with two signs that are the **same**, the answer is **positive**.

$$-3 \times 6 = -18 \qquad 7 \times (-2) = -14 \qquad 24 \div (-6) = -4$$

$$-5 \times (-7) = 35 \qquad -100 \div (-5) = 20$$

Key Point

$- \times - = + \qquad - \times + = -$

$+ \times + = + \qquad + \times - = -$

Calculating with Decimals

- When **adding** or **subtracting** decimals, keep the decimal points under each other.

Key Point

The number of decimal places in the answer is the same as the total number of decimal places in the original calculation.

Two parcels weigh 6.3kg and 11.27kg. What is their total weight?

$$\begin{array}{r} 6.30 \\ 11.27\,+ \\ \hline 17.57 \end{array}$$

Total weight: 17.57kg

Remember 6.3 is the same as 6.30

The decimal point in the answer should be in line with the other decimal points.

- When **multiplying**, remove the decimal points first.

Two figures after the decimal points in the question, so two figures after the decimal point in the answer.

Work out 0.6×0.2

$6 \times 2 = 12$

$0.6 \times 0.2 = 0.12$

Work out 1.46×2.3

$146 \times 23 = 3358$

$1.46 \times 2.3 = 3.358$

Three figures after the decimal points on both sides of the equals sign.

- When **dividing**, always divide by a whole number.

In order to divide by a whole number, multiply everything by 10.

Work out $27.632 \div 0.2$

$276.32 \div 2 = 138.16$

Work out $36.14 \div 0.002$

$36\,140 \div 2 = 18\,070$

Multiply everything by 1000.

Standard Form

- **Standard form** is used to represent very small or very large numbers.
- A number in standard form is written in the form $A \times 10^{n}$, where $1 \leqslant A < 10$ and n is an **integer**.
- For numbers less than 1, n is negative.

$n = 5$ since the decimal point has to move 5 places to the right to go from 7.23 to 723 000.

Write the following numbers in standard form:

a) 723 000 $723\,000 = 7.23 \times 10^{5}$

b) 0.000 006 3 $0.000\,006\,3 = 6.3 \times 10^{-6}$

$n = -6$ since the decimal point has to move 6 places to the left to go from 6.3 to 0.000 006 3.

Quick Test

1. a) -7×-3 b) $-6 - 7$
2. Write 0.001 63 in standard form.
3. Put in order, smallest to largest: 220, −220, −2201, 1022, 2200
4. Vishnaal spent £7.84.
   How much change did he receive from £20?

Key Words

place value
negative number
positive number
standard form
integer

Number 3

You must be able to:

- Use the ideas of prime numbers, factors (divisors), multiples, highest common factor, lowest common multiple and prime factors
- Show prime factor decomposition
- Use systematic listing strategies, including the product rule for counting.

Types of Number

- **Multiples** are found in the 'times table' of the number, e.g.
  Multiples of 4 = {4, 8, 12, 16 ...}
- A **factor** (divisor) is a number that will divide exactly into another number, e.g.
  Factors of 12 = {1, 2, 3, 4, 6, 12}
- A **prime number** has only two factors: itself and 1.
- **Square numbers** are the results of multiplying together two numbers that are the same. They are shown using a **power** of 2, e.g.

  $1 \times 1 = 1$ $2 \times 2 = 4$ $3 \times 3 = 9$ $4 \times 4 = 16$

  $1^2 = 1$ $2^2 = 4$ $3^2 = 9$ $4^2 = 16$

- A **square root** is the **inverse** (opposite) of a square, e.g.
  $6^2 = 36$ $\sqrt{36} = 6$
- **Cube numbers** are the results of multiplying together three numbers that are the same. They are shown using a power of 3, e.g.

  $1 \times 1 \times 1 = 1$ $2 \times 2 \times 2 = 8$ $3 \times 3 \times 3 = 27$ $4 \times 4 \times 4 = 64$

  $1^3 = 1$ $2^3 = 8$ $3^3 = 27$ $4^3 = 64$

- A **cube root** is the inverse of a cube, e.g.
  $5^3 = 125$ $\sqrt[3]{125} = 5$

> **Key Point**
>
> When listing factors, look for pairs of numbers, then you will not forget any.

> **Key Point**
>
> You are expected to know the squares and square roots up to $15 \times 15 = 225$ and that $\sqrt[3]{1} = 1$, $\sqrt[3]{8} = 2$, $\sqrt[3]{27} = 3$, $\sqrt[3]{64} = 4$, $\sqrt[3]{125} = 5$ and $\sqrt[3]{1000} = 10$.

Prime Factors, LCM and HCF

- The **lowest common multiple (LCM)** of two numbers is the lowest integer that is a multiple of both numbers.

  > Find the LCM of 12 and 18.
  >
  > Multiples of 12 = {12, 24, ⓐ36, 48 ...}
  >
  > Multiples of 18 = {18, ⓐ36, 54 ...}
  >
  > 36 is the smallest number that is in both lists, so 36 is the LCM.

  Write out the multiples of 12 and 18 until you get a common value.

- The **highest common factor (HCF)** of two numbers is the largest integer that will divide exactly into both numbers.

  > Find the HCF of 45 and 60.
  >
  > Factors of 45 = {1, 3, 5, 9, ⓐ15, 45}
  >
  > Factors of 60 = {1, 2, 3, 4, 5, 6, 10, 12, ⓐ15, 20, 30, 60}
  >
  > 15 is the largest number that is in both lists, so 15 is the HCF.

  Write out the factors of 45 and 60 and look for the highest common value.

- Factors of a number that are also prime numbers are called prime factors.

Write 48 as a product of prime factors.

Method 1:
Prime Factor Decomposition

Method 2:
A Factor Tree

There are two ways of attempting this question.

Keep splitting values into factors until the ends of all branches are prime numbers.

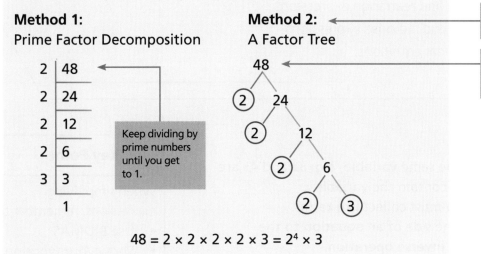

Keep dividing by prime numbers until you get to 1.

$$48 = 2 \times 2 \times 2 \times 2 \times 3 = 2^4 \times 3$$

Choices and Outcomes

- The **product rule** (multiplication rule) for counting is used to find the total number of combinations possible for a given scenario.
- The product rule states: if there are A ways of doing Task 1 and B ways of doing Task 2, then there are $A \times B$ ways of doing both tasks.

Key Point

Always list numbers in order, smallest to largest, unless told otherwise.

In a restaurant there are three different starters (X, Y, Z) and six different main meals (1, 2, 3, 4, 5, 6) to choose from.

How many possible combinations are there when choosing a starter and a main meal?

Combinations are: X1, X2, X3, X4, X5, X6, Y1, Y2, Y3, Y4, Y5, Y6, Z1, Z2, Z3, Z4, Z5, Z6 (18 combinations)

Using the product rule:
3 (starters) × 6 (mains) = 18 (combinations)

Key Words

multiple
factor
prime number
square number
power
square root
inverse
cube number
cube root
lowest common multiple (LCM)
highest common factor (HCF)
prime factor
product rule

Quick Test

1. Express 36 as a product of prime factors using any method.
2. Find **a)** the HCF and **b)** the LCM of 12 and 15.
3. The number 88 can be written as $2^m \times p$ where m and p are prime numbers. Work out the values of m and p.
4. Red buses leave the bus garage every 12 minutes.
   Blue buses leave the bus garage every 20 minutes.
   A red bus and a blue bus both leave the garage at 9am.
   At what time will a red bus and a blue bus next leave the garage together?

Basic Algebra

You must be able to:

- Use and understand algebraic notation and vocabulary
- Simplify and rearrange expressions
- Expand and factorise expressions
- Solve linear equations.

Basic Algebra

- 'Like terms' are **terms** with the same **variable**, e.g. $3x$ and $4x$ are like terms because they both contain the variable x.
- To simplify an **expression** you must collect like terms.
- When moving a term from one side of an **equation** to the other, you must carry out the **inverse operation**.

> **Key Point**
>
> When simplifying expressions, remember to:
> - Use BIDMAS
> - Show your working.

Simplify $3x + 3y - 7x + y$

$4y - 4x$ $3x - 7x = -4x$
$3y + y = 4y$

Simplify $9p^2 + 7p - qp + pq - p^2$ $qp = pq$

$8p^2 + 7p$ $9p^2 - p^2 = 8p^2$
$-qp + pq = 0$

- Substitution means replacing variables with numbers.

> **Key Point**
>
> An expression does not contain an = sign.

If $y = 4$ and $t = 6$, work out the value of $7y - 6t$

$$7y - 6t = 7 \times 4 - 6 \times 6$$
$$= 28 - 36$$
$$= -8$$

If $q = 5$, $r = 2$ and $z = -3$, work out the value of $rq + z^2$

Use brackets as the minus sign is also squared.

$$rq + z^2 = 2 \times 5 + (-3)^2$$
$$= 10 + 9$$
$$= 19$$

> **Key Point**
>
> Always apply the rules:
> $- \times - = +$ $- \times + = -$
> $+ \times + = +$ $+ \times - = -$

- To expand (multiply out) brackets, every term in the bracket is multiplied by the term outside the bracket.

Expand $3(x + 2)$ Expand $5p(p - 2)$ $3 \times x = 3x$ and $3 \times 2 = +6$

$3x + 6$ $5p^2 - 10p$ $5p \times p = 5p^2$
$5p \times (-2) = -10p$

Expand and simplify $4y(2y - 3) - 3y(y - 2)$

$8y^2 - 12y - 3y^2 + 6y$ Note that $-3y \times -2 = +6y$
$= 5y^2 - 6y$

Factorisation

- **Factorisation** is the reverse of expanding brackets, i.e. you take out a common factor and put brackets into the expression.
- To factorise, you should look for common factors in every term.

Factorise $12x + 4$ ← 4 is the HCF of 12 and 4.

$4(3x + 1)$

Factorise $3x^2 - 6x$ Factorise $3p^3 - 2p^2 + 8p$

$3x(x - 2)$ ← Remember $x^2 = x \times x$ $p(3p^2 - 2p + 8)$

> **Key Point**
>
> To factorise completely, always take out the highest common factor, e.g. 3 is the HCF of 3 and 6.

Linear Equations

- A **linear equation** does not contain any variables with a power greater than 1.
- When you solve an equation, you are finding an unknown number, represented by a letter, e.g. x.

Solve $5x + 6 = 16$ Solve $4(x - 2) = 20$

$5x = 10$ ← The inverse of + is − $4x - 8 = 20$ ← Expand the brackets.

$x = 2$ ← The inverse of × is ÷ $4x = 28$

          $x = 7$

> **Key Point**
>
> When moving a term from one side of an equation to the other, you must carry out the inverse operation.

Solve $5y - 4 = 3y + 10$ Solve $\frac{5s}{4} + 3 = 18$

$5y - 3y = 10 + 4$ $\frac{5s}{4} = 15$ ← Subtract 3 from both sides.

$2y = 14$ ← Collect all the letter terms on one side. $5s = 60$ ← Multiply both sides by 4.

$y = 7$ $s = 12$ ← Divide both sides by 5.

- Equations can be used to represent real-life problems.
- The equation should be rearranged to solve the problem.

Mary bought nine candles. She used a £3 gift voucher as part payment. The balance left to pay was £5.55

What was the cost of one candle (c)?   Use the information given to set up an equation.

$9c - 3 = 5.55$ Solve to find the cost of one candle.

$9c = 8.55$

$c = \frac{8.55}{9} = £0.95$ or 95p

> **Key Words**
>
> term
> variable
> expression
> equation
> inverse operation
> factorisation
> linear equation

> **Quick Test**
>
> 1. Factorise $5x + 10$
> 2. Solve $7x - 2 = 12$
> 3. Simplify $2y - 7 + 4y + 2$
> 4. Work out the value of $3p^3 - 7q$, when $p = -4$ and $q = -3$.
> 5. Expand the following expression: $3t(4t - 1)$

Factorisation and Formulae

You must be able to:

- Expand the product of two binomials
- Factorise a quadratic expression
- Understand and use formulae
- Rearrange and change the subject of a formula.

Binomial Expansion

- A **binomial** is an **expression** that contains two terms, e.g. $x + 2$ or $3y - 4$.
- The product of two binomials is obtained when they are multiplied together, e.g. $(2r + 7)(3r - 6)$.
- To **expand** (or multiply out) the brackets, every term in the first set of brackets must be multiplied by every term in the second set of brackets.

Expand and simplify $(x + 2)(x + 6)$

×	x	+2
x	x^2	+2x
+6	+6x	+12

$x^2 + 2x + 6x + 12$
$= x^2 + 8x + 12$ ←

Simplify by collecting like terms.

Expand and simplify $(2y + 4)(3y - 2)$

×	2y	+4
3y	$6y^2$	+12y
−2	−4y	−8

$6y^2 + 12y - 4y - 8$
$= 6y^2 + 8y - 8$

> **Key Point**
>
> Take care over + and − signs.

Quadratic Factorisation

- An expression that contains a squared term is called **quadratic**.
- Some quadratic expressions can be written as a product of two binomials.
- When written in factorised form, the new expression is equivalent to the original quadratic.

> **Key Point**
>
> Check you have factorised correctly by expanding the brackets – your expressions should be equivalent.

Factorise $x^2 + 5x + 6$

$x^2 + 5x + 6$ ←

$(x + 2)(x + 3)$ ←

Find a pair of numbers with a sum of +5 and a product of +6.

$(+2) + (+3) = +5$ and $(+2) \times (+3) = +6$

Factorise the expression $x^2 - 4x + 3$

×	x	-1
x	x^2	$-x$
-3	$-3x$	$+3$

$(x - 1)(x - 3)$

The missing terms need to have a product of $+3$ and a sum of -4, i.e. -1 and -3.

Write the expression as a product: the **first row** gives you the **first bracket** and the first column gives you the **second bracket**.

Changing the Subject of a Formula

- A **formula** is a way of describing a rule or fact.
- A formula is written as an algebraic equation.
- The **subject** of a formula appears once on the left-hand side.
- To change the subject, a formula must be rearranged using **inverse operations**.

Make p the subject of $5p - 7 = r$

$5p = r + 7$

$p = \dfrac{r + 7}{5}$

> **Key Point**
>
> When rearranging formulae remember to use inverse operations. Finish by writing the formula out with the new subject on the left-hand side.

This formula can be used to change temperature in degrees Fahrenheit to temperature in degrees Celsius:
$C = \frac{5}{9}(F - 32)$

In Iceland, the lowest recorded temperature on a certain day is $-20°C$.

What is this temperature in degrees Fahrenheit?

$-20 = \frac{5}{9}(F - 32)$

$-180 = 5(F - 32)$

$-36 = F - 32$

$F = -4°F$

The formula must be rearranged to find the value of F.

The answer is -4 degrees Fahrenheit.

Make r the subject of the formula $P = 3(r - 1)$

$\dfrac{P}{3} = r - 1$

$r = \dfrac{P}{3} + 1$

The formula for calculating the area of a circle is $A = \pi r^2$.

Make r the subject.

$\dfrac{A}{\pi} = r^2$

$r = \sqrt{\dfrac{A}{\pi}}$

π can be treated as a numerical term.

Only the positive root is needed as r is a length.

> **Key Words**
>
> binomial
> expression
> expand
> quadratic
> formula
> subject
> inverse operation

> **Quick Test**
>
> 1. $T = 30w + 20$. Work out the value of w when $T = 290$.
> 2. Factorise $x^2 + 8x + 7$
> 3. Make q the subject of $6q - 5 = 2t$
> 4. Make y the subject of the formula $\dfrac{x + 2}{3} = 2(y - 1)$

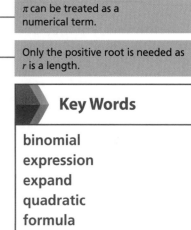

Ratio and Proportion

You must be able to:

- Use ratio notation and reduce ratios to their simplest form
- Divide quantities into given ratios
- Apply ratio to real contexts and problems, including best buys
- Solve problems using direct and inverse proportion.

Ratio

- **Ratios** are used to compare quantities.
- You can **simplify** a ratio. This is like cancelling down a fraction.

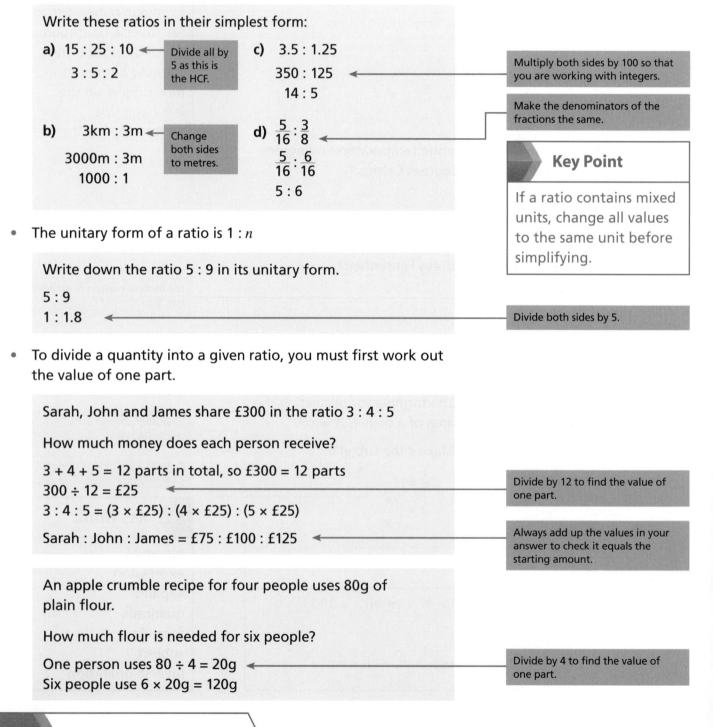

Write these ratios in their simplest form:

a) 15 : 25 : 10 ← Divide all by 5 as this is the HCF.

3 : 5 : 2

b) 3km : 3m ← Change both sides to metres.

3000m : 3m

1000 : 1

c) 3.5 : 1.25

350 : 125 ← Multiply both sides by 100 so that you are working with integers.

14 : 5

d) $\frac{5}{16} : \frac{3}{8}$ ← Make the denominators of the fractions the same.

$\frac{5}{16} : \frac{6}{16}$

5 : 6

Key Point

If a ratio contains mixed units, change all values to the same unit before simplifying.

- The unitary form of a ratio is 1 : n

Write down the ratio 5 : 9 in its unitary form.

5 : 9

1 : 1.8 ← Divide both sides by 5.

- To divide a quantity into a given ratio, you must first work out the value of one part.

Sarah, John and James share £300 in the ratio 3 : 4 : 5

How much money does each person receive?

3 + 4 + 5 = 12 parts in total, so £300 = 12 parts

300 ÷ 12 = £25 ← Divide by 12 to find the value of one part.

3 : 4 : 5 = (3 × £25) : (4 × £25) : (5 × £25)

Sarah : John : James = £75 : £100 : £125 ← Always add up the values in your answer to check it equals the starting amount.

An apple crumble recipe for four people uses 80g of plain flour.

How much flour is needed for six people?

One person uses 80 ÷ 4 = 20g ← Divide by 4 to find the value of one part.

Six people use 6 × 20g = 120g

Best Buys

- Working out the value of one part allows you to make like-for-like comparisons.

> Teabags are sold in two different sized packs:
>
> Pack A holds 80 teabags and costs £1.80.
>
> Pack B holds 200 teabags and costs £3.80.
>
> Which is the best buy?
>
> Pack A   180 ÷ 80 = 2.25p ◄──── | Work out the cost of one teabag in each pack.
>
> Pack B   380 ÷ 200 = 1.9p
>
> Packet B is the best buy (the cheaper price per teabag).

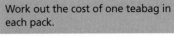

Direct Proportion Problems

- Quantities are in **direct proportion** if their ratio remains the same as they are increased or decreased.
- Direct proportion uses the symbol α.
- $y \propto x$ means y is directly proportional to x or $y = kx$.
- k is called the **constant of proportionality**.
- Quantities can also be in **inverse proportion**.
- This is shown by $y \propto \frac{1}{x}$ or $y = \frac{k}{x}$.

> Mass is directly proportional to volume.
> When mass = 100g, then volume = 25cm³.
>
> Work out the mass for a volume of 40cm³.
>
> **Stage 1**
>
> $m \propto v$
>
> $m = kv$ ◄──── | Substitute in the values for m (mass) and v (volume).
>
> $100 = k \times 25$
>
> $k = 4$ ◄──── | You now have a value for k.
>
> **Stage 2**
>
> $m = kv$ ◄──── | Substitute in the values for k and v (volume).
>
> $m = 4 \times 40$
>
> mass = 160g

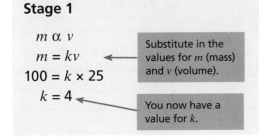

Key Point

If quantities are in inverse proportion, as x increases, y decreases.

The graph produced is a curve.

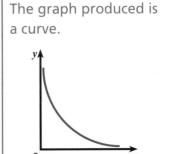

Quick Test

1. £180 is shared in the ratio 2 : 3
   What is value of the larger share?
2. Simplify 3 hours : 45 minutes
3. a) Cat food comes in two different sized boxes:
      Box A has 12 sachets and costs £3.80.
      Box B has 45 sachets and costs £14.00.
      Which is the best buy? Explain your answer clearly.
   b) Tiggles the cat eats three sachets a day. How many of the best buy boxes have to be bought to feed Tiggles in June?

Key Words

ratio
simplify
direct proportion
constant of
   proportionality
inverse proportion

Variation and Compound Measures

You must be able to:

- Use compound units and solve problems involving compound measures
- Calculate compound interest
- Understand graphs that illustrate direct and inverse proportion.

Compound Measures

- A **compound measure** is a measure that involves two or more other measures.
- **Speed** is a compound measure – it involves measures of distance and time.

LEARN

$$\text{Speed}\ (S) = \frac{\text{Total Distance}\ (D)}{\text{Time Taken}\ (T)} \qquad D = S \times T \qquad T = \frac{D}{S}$$

> **Key Point**
>
> Units for speed include kilometres per hour (km/h) or metres per second (m/s).

A snail crawls a distance of 30cm in 3 minutes 20 seconds.

Work out its speed in m/s.

30cm = 0.3m ←

3 minutes 20 seconds = 200 seconds ←   Change the total time into seconds.

Speed = $\frac{\text{Distance}}{\text{Time}}$

$= \frac{0.3}{200}$

$= 0.0015\text{m/s}$

> Change 30cm into metres by dividing by 100 (100cm = 1m).

> **Key Point**
>
> Always check that the units in the question are the same as the units required in the answer.

- **Density** and **pressure** are compound measures.

LEARN

$$\text{Density}\ (D) = \frac{\text{Mass}\ (M)}{\text{Volume}\ (V)} \qquad M = D \times V \qquad V = \frac{M}{D}$$

LEARN

$$\text{Pressure}\ (P) = \frac{\text{Force}\ (F)}{\text{Area}\ (A)} \qquad F = P \times A \qquad A = \frac{F}{P}$$

Compound Interest and Repeated Percentage Change

- **Compound interest** is calculated based on the total amount of money invested, plus any interest previously earned.
- This formula can be used to calculate how money invested grows with time:

> **Key Point**
>
> Units for density include kg/m³ or g/cm³.
>
> Pressure is expressed in newtons (N) per square metre (m²). A force of 1N applied to 1m² is called 1 pascal (Pa).

LEARN

$$\text{Final Amount}\ (A) = \text{Original Amount} \times \left(1 + \frac{\text{Rate}}{100}\right)^{\text{time}}$$

- To calculate **depreciation** (loss in value), the plus sign in the formula is changed to a minus sign.

> **Key Point**
>
> Simple interest is when the interest is paid out rather than being added to the money invested.

£3000 is invested at 2% compound interest per annum.

Calculate how much money there will be after four years.
Give your answer to the nearest pound.

$A = 3000 \times \left(1 + \dfrac{2}{100}\right)^4$

$= 3000 \times (1.02)^4$

$= 3000 \times 1.0824 = £3247$ (to the nearest £)

Use the formula:
$A = \text{Original Amount} \times \left(1 + \dfrac{\text{Rate}}{100}\right)^{time}$

The value of a new car is £8000.
The car depreciates in value by 10% each year.

Work out the car's value after five years.

$A = 8000 \times \left(1 - \dfrac{10}{100}\right)^5$

$= 8000 \times (0.9)^5$

$= 8000 \times 0.5905 = £4723.92$

Use the formula:
$A = \text{Original Amount} \times \left(1 - \dfrac{\text{Rate}}{100}\right)^{time}$

Direct Proportion

- In a graph, direct proportion can be represented by a straight line that passes through the origin (0, 0).

This table gives information about the journey of a car during a half-hour time period.

Distance (miles)	0	6	12	18
Time (mins)	0	10	20	30

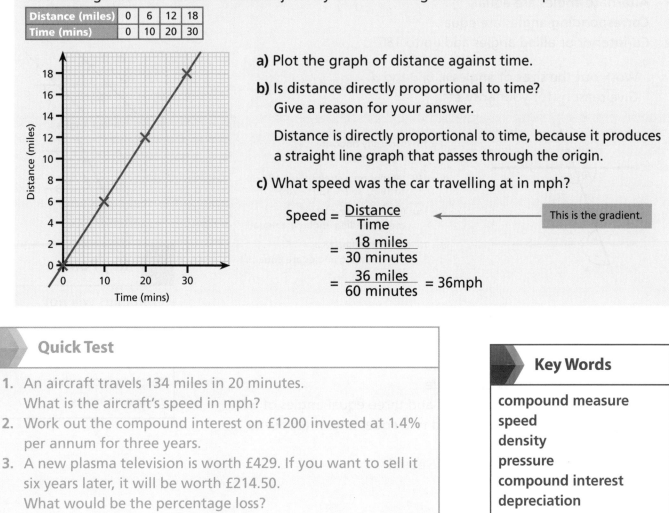

a) Plot the graph of distance against time.

b) Is distance directly proportional to time?
Give a reason for your answer.

Distance is directly proportional to time, because it produces a straight line graph that passes through the origin.

c) What speed was the car travelling at in mph?

$\text{Speed} = \dfrac{\text{Distance}}{\text{Time}}$

This is the gradient.

$= \dfrac{18 \text{ miles}}{30 \text{ minutes}}$

$= \dfrac{36 \text{ miles}}{60 \text{ minutes}} = 36 \text{mph}$

Quick Test

1. An aircraft travels 134 miles in 20 minutes.
   What is the aircraft's speed in mph?
2. Work out the compound interest on £1200 invested at 1.4% per annum for three years.
3. A new plasma television is worth £429. If you want to sell it six years later, it will be worth £214.50.
   What would be the percentage loss?

Key Words

compound measure
speed
density
pressure
compound interest
depreciation

Angles and Shapes 1

You must be able to:

- Recognise relationships between angles
- Use the properties of angles to work out unknown angles
- Recognise different types of triangle
- Understand and use the properties of special types of quadrilaterals.

Angle Facts

- There are three types of angle:
  - **acute:** less than 90°
  - **obtuse:** between 90° and 180°
  - **reflex:** between 180° and 360°.
- Angles on a straight line add up to 180°.
- Angles around a point add to 360°.
- **Vertically opposite** angles are equal.

Angles in Parallel Lines

- Parallel lines never meet. The lines are always the same distance apart.
- **Alternate** angles are equal.
- **Corresponding** angles are equal.
- Co-interior or **allied** angles add up to 180°.

Work out the sizes of angles a, b, c and d.
Give reasons for your answers.

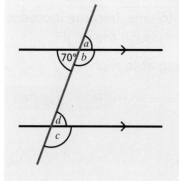

$a = 70°$ (vertically opposite angles are equal)

$b = 110°$ (angles on a straight line add up to 180°, so $b = 180° - 70°$)

$c = 110°$ (corresponding to b; corresponding angles are equal)

$d = 70°$ (corresponding to a; corresponding angles are equal)

Triangles

- Angles in a triangle add up to 180°.
- There are several types of triangle:
  - **equilateral:** three equal sides and three equal angles of 60°
  - **isosceles:** two equal sides and two equal angles (opposite the equal sides)
  - **scalene:** no sides or angles are equal
  - **right-angled:** one 90° angle.

Alternate Angles

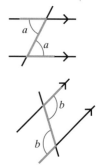

Corresponding Angles

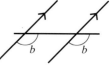

Allied Angles

$c + d = 180°$

Key Point

Examiners will **not** accept terms like 'Z angles' or 'F angles'. Always use correct terminology when giving reasons.

ABC is an isosceles triangle and HE is parallel to GD. BAF is a straight line. Angle $FAE = 81°$

Calculate **a)** angle ABC and **b)** angle ACB.
Give reasons for your answers.

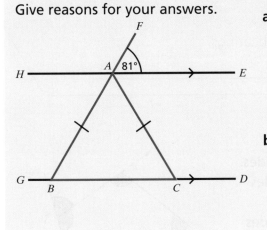

There are several different ways of solving this question.

a) Angle $HAB =$ 81° (vertically opposite FAE), so angle $ABC = 81°$ (alternate angle to HAB)

b) Angle $ACB = 81°$ (angle $ABC =$ angle ACB; base angles of an isosceles triangle are equal.)

Special Quadrilaterals

- The interior angles in a quadrilateral add up to 360°.
- The order of rotational symmetry is the number of times a shape looks the same when it is rotated 360° (one full turn).
- You need to know the properties of these special quadrilaterals:

	Sides	Angles	Lines of Symmetry	Rotational Symmetry	Diagonals
parallelogram	opposite sides are equal and parallel	diagonally opposite angles are equal	none	order 2	diagonals bisect each other
rhombus	all sides are equal and opposite sides are parallel	opposite angles are equal	two	order 2	diagonals bisect each other at 90°
kite	two pairs of adjacent sides are equal	one pair of opposite angles are equal	one	none	diagonals cross at 90°
trapezium	one pair of opposite sides are parallel		none (an isosceles trapezium has one)	none	

Quick Test

1. Name all the quadrilaterals that can be drawn with lines of lengths:
   a) 4cm, 7cm, 4cm, 7cm   **b)** 6cm, 6cm, 6cm, 6cm
2. $EFGH$ is a trapezium with EH parallel to FG. FE and GH are produced (made longer) to meet at J. Angle $EHF = 62°$, angle $EFH = 25°$ and angle $JGF = 77°$. Calculate the size of angle EJH.

Key Words

acute	equilateral
obtuse	isosceles
reflex	scalene
vertically opposite	right-angled
	parallelogram
alternate	rhombus
corresponding	kite
allied	trapezium

Angles and Shapes 2

You must be able to:

- Work out angles in a polygon
- Answer questions on regular polygons
- Understand scale drawings and use bearings.

Angles in a Polygon

- A **polygon** is a closed shape with at least three straight sides.
- **Regular** polygons are shapes where all the sides and angles are equal.
- **Irregular** polygons are shapes where some or all of the sides and angles are different.
- For all polygons:
  - at any **vertex** (corner): **interior** angle + **exterior** angle = 180°
  - sum of all exterior angles = 360°.
- To work out the sum of the interior angles in a polygon, you can split it into triangles from one vertex.
- For example, a pentagon can be divided into three triangles, so the sum of the interior angles is 3 × 180° = 540°.
- The sum of the interior angles for any polygon can be calculated using the formula:

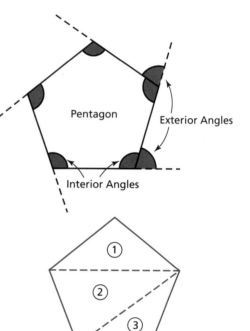

Pentagon

Exterior Angles

Interior Angles

> **LEARN**
> Sum = $(n - 2) \times 180°$ Where n = number of sides

Work out the sum of the interior angles of a decagon (10 sides).

Sum = (10 − 2) × 180°
$\quad$ = 8 × 180°
$\quad$ = 1440°

> Use the formula:
> Sum = $(n - 2) \times 180°$

Regular Polygons

- In regular polygons:

> **LEARN**
> Number of Sides (n) × Exterior Angle = 360°
> So, Exterior Angle = 360° ÷ n

Work out the size of the interior angles in a regular hexagon (six sides).

Exterior angle $\quad$ = 360° ÷ 6 = 60°
Interior angle + 60° = 180°
Interior angle $\qquad$ = 180° − 60°
$\qquad\qquad\qquad$ = 120°

> Use the formula:
> Exterior Angle = 360° ÷ n

> Interior Angle + Exterior Angle = 180°

A regular polygon has an interior angle of 156°.

Work out the number of sides that the polygon has.

Exterior angle = 180° – interior angle
= 180° – 156° = 24°

Number of sides = 360° ÷ 24° = 15

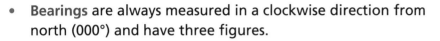

156°

Scale Drawings and Bearings

- **Bearings** are always measured in a clockwise direction from north (000°) and have three figures.

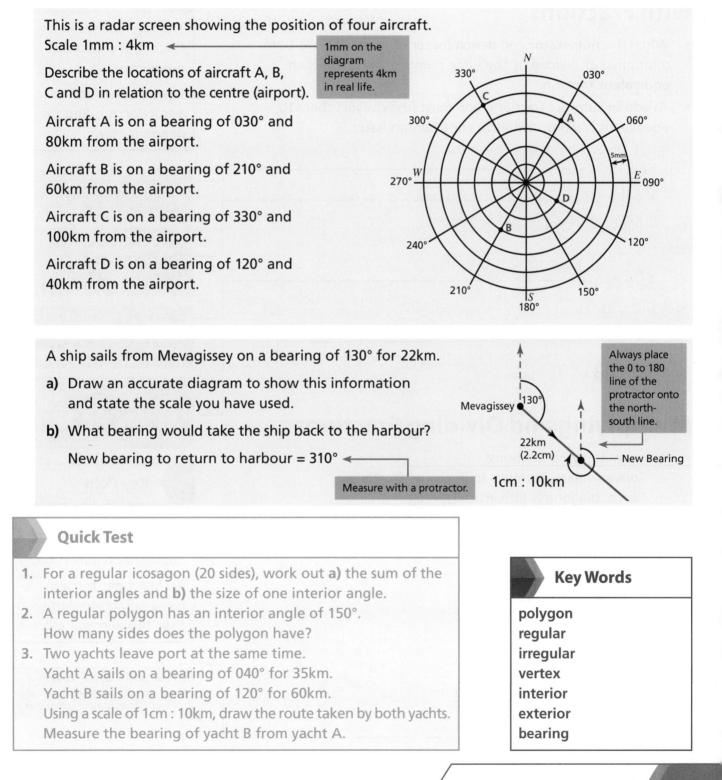

This is a radar screen showing the position of four aircraft.
Scale 1mm : 4km

1mm on the diagram represents 4km in real life.

Describe the locations of aircraft A, B, C and D in relation to the centre (airport).

Aircraft A is on a bearing of 030° and 80km from the airport.

Aircraft B is on a bearing of 210° and 60km from the airport.

Aircraft C is on a bearing of 330° and 100km from the airport.

Aircraft D is on a bearing of 120° and 40km from the airport.

A ship sails from Mevagissey on a bearing of 130° for 22km.

a) Draw an accurate diagram to show this information and state the scale you have used.

b) What bearing would take the ship back to the harbour?

New bearing to return to harbour = 310°

Measure with a protractor.

Always place the 0 to 180 line of the protractor onto the north-south line.

Mevagissey 130°

22km (2.2cm)

New Bearing

1cm : 10km

Quick Test

1. For a regular icosagon (20 sides), work out **a)** the sum of the interior angles and **b)** the size of one interior angle.
2. A regular polygon has an interior angle of 150°. How many sides does the polygon have?
3. Two yachts leave port at the same time.
   Yacht A sails on a bearing of 040° for 35km.
   Yacht B sails on a bearing of 120° for 60km.
   Using a scale of 1cm : 10km, draw the route taken by both yachts.
   Measure the bearing of yacht B from yacht A.

Key Words

polygon
regular
irregular
vertex
interior
exterior
bearing

Fractions

You must be able to:

- Add, subtract, multiply and divide fractions
- Change terminating decimals into corresponding fractions and vice versa
- Express one quantity as a fraction of another
- Use a calculator to change a fraction to a decimal.

Adding, Subtracting and Calculating with Fractions

- When the **numerator** and **denominator** of a fraction are both multiplied or divided by the same number, the result is an **equivalent** fraction.
- To add or subtract fractions, you must first convert them to equivalent fractions with the same denominator.

$$5\frac{3}{5} + 1\frac{2}{7}$$

$$= 6\frac{21}{35} + \frac{10}{35}$$

$$= 6\frac{31}{35}$$

$$4\frac{1}{4} - 2\frac{3}{5}$$

$$= \frac{17}{4} - \frac{13}{5}$$

$$= \frac{85}{20} - \frac{52}{20}$$

$$= \frac{33}{20} = 1\frac{13}{20}$$

> **Key Point**
>
> The top number in a fraction is called the **numerator**.
>
> The bottom number in a fraction is called the **denominator**.

Add whole numbers and then change fractions so that they have the same denominator.

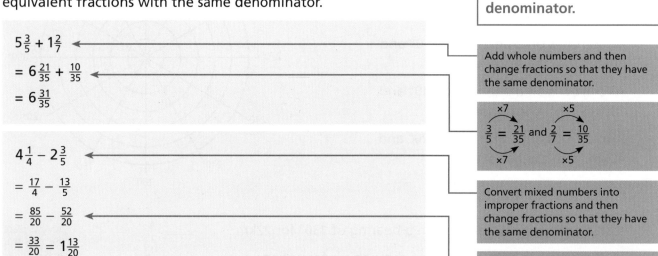

$$\frac{3}{5} = \frac{21}{35} \text{ and } \frac{2}{7} = \frac{10}{35}$$ (×7 and ×5)

Convert mixed numbers into improper fractions and then change fractions so that they have the same denominator.

$$\frac{17}{4} = \frac{85}{20} \text{ and } \frac{13}{5} = \frac{52}{20}$$ (×5 and ×4)

Multiplying and Dividing Fractions

- When multiplying fractions:
  - Convert mixed numbers to improper fractions
  - Cancel diagonally (or vertically) only
  - Then multiply the numerators together and the denominators together.
- When dividing fractions:
  - Invert the second fraction
  - Change the ÷ sign to a × sign
  - Then multiply.

$$6\frac{4}{5} \times 2\frac{1}{2}$$

$$= \frac{^{17}\cancel{34}}{_1\cancel{5}} \times \frac{^1\cancel{5}}{_1\cancel{2}} = \frac{17}{1} = 17$$

$$\frac{1}{4} \div \frac{7}{10}$$

$$= \frac{1}{_2\cancel{4}} \times \frac{\cancel{10}^5}{7} = \frac{5}{14}$$

> **Key Point**
>
> A **mixed number** contains a whole number and a fraction.
>
> An **improper fraction** has a numerator larger than the denominator.

Cancel down and then multiply. If the answer is an improper fraction, convert it back to a mixed number.

 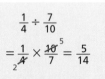

Rational Numbers, Reciprocals, and Terminating and Recurring Decimals

- **Rational numbers** are numbers that **can be written exactly** as a fraction or decimal, e.g. $\frac{1}{4} = 0.25$
- π is an **irrational number** as it would continue forever as a decimal $3.141\,592\ldots$
- The **reciprocal** of a number (n) is 1 divided by the number, i.e. $\frac{1}{n}$
- A **terminating decimal** is a decimal with a finite number of digits (it ends), e.g. 0.75, 0.36
- A **recurring** decimal has a digit or group of digits that repeat indefinitely. It is shown with a dot over the recurring digit(s), e.g.

$0.1\dot{6} = 0.1666666\ldots$
$0.\dot{3}\dot{7} = 0.37373737\ldots$

Key Point

To change $\frac{3}{5}$ to a decimal on a calculator, press

| ON | 3 | ÷ | 5 | = |

Use $\boxed{S \Leftrightarrow D}$ to change the answer between a decimal and a fraction.

Change $\frac{3}{5}$ into a decimal.

$\frac{3}{5} = 3.00 \div 5$
$= 0.60$

Change the following decimals into fractions. Give your answers in their simplest form.

a) 0.7 $\frac{7}{10}$

b) 0.35 $\frac{35}{100} = \frac{7}{20}$

c) 0.417 $\frac{417}{1000}$

One Quantity as a Fraction of Another

Express 20 minutes as a fraction of 3 hours 20 minutes.

3 hours 20 minutes = 200 minutes ← Convert to minutes.

20 minutes as a fraction of 200 minutes $= \frac{20}{200} = \frac{1}{10}$ ← Cancel down.

A school has 712 students.
$\frac{5}{8}$ of the students travel to school by bus and the rest walk.

How many students walk to school?

Method 1
712 ÷ 8 = 89
89 × 5 = 445

Number who walk = 712 − 445
                = 267 students

Method 2
$1 - \frac{5}{8} = \frac{3}{8}$

$\frac{3}{8} \times 712 = 267$ students

To find the number of students who take the bus, work out $\frac{1}{8}$ of 712 and then multiply by 5 to get $\frac{5}{8}$.

Start by working out how many students walk to school as a fraction.

Key Words

numerator
denominator
equivalent
rational number
irrational number
reciprocal
terminating decimal
recurring

Quick Test

1. Which is larger: $\frac{3}{4} + \frac{1}{5}$ or $\frac{3}{4} \div \frac{1}{5}$?
   You must show your working.
2. a) Use your calculator to change $\frac{1}{9}$ to a decimal.
   b) Is the resulting decimal terminating or recurring?
3. Express 15cm as a fraction of 4m in its simplest form.

Percentages 1

You must be able to:

- Convert percentages to fractions and decimals and vice versa
- Calculate a percentage of a quantity
- Work out percentage change.

Converting Between Fractions, Decimals and Percentages

- To change **from a** percentage to a fraction or a decimal, **divide** by 100.

Change 35% to:

a) a fraction.

$35\% = \frac{35}{100} = \frac{7}{20}$

b) a decimal.

$35\% = 0.35$

- To change from a fraction or a decimal **to a percentage**, **multiply** by 100.

a) Convert $\frac{2}{5}$ to a percentage.

$\frac{2}{5} \times 100 = \frac{200}{5} = 40\%$

b) Convert 0.567 to a percentage.

$0.567 = 56.7\%$

> **Key Point**
>
> Percentages to learn:
> $100\% = 1$ (the whole)
> $75\% = \frac{3}{4} = 0.75$
> $50\% = \frac{1}{2} = 0.5$
> $25\% = \frac{1}{4} = 0.25$
> $10\% = \frac{1}{10} = 0.1$

Move the decimal point two places to the **left** when **dividing** by 100.

Move the decimal point two places to the **right** when **multiplying** by 100.

A Percentage of a Quantity

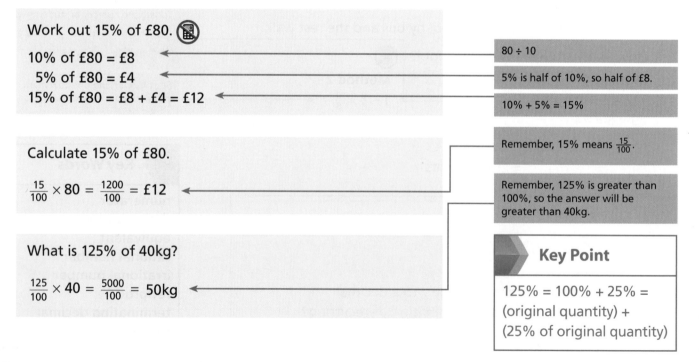

Work out 15% of £80.

10% of £80 = £8

5% of £80 = £4

15% of £80 = £8 + £4 = £12

$80 \div 10$

5% is half of 10%, so half of £8.

10% + 5% = 15%

Calculate 15% of £80.

$\frac{15}{100} \times 80 = \frac{1200}{100} = £12$

Remember, 15% means $\frac{15}{100}$.

Remember, 125% is greater than 100%, so the answer will be greater than 40kg.

What is 125% of 40kg?

$\frac{125}{100} \times 40 = \frac{5000}{100} = 50\text{kg}$

> **Key Point**
>
> $125\% = 100\% + 25\% =$ (original quantity) + (25% of original quantity)

Percentage Change

- To calculate percentage change, work out the change in value, e.g. the increase or decrease, and use the formula:

LEARN

$$\text{Percentage Change} = \frac{\text{Change in Value}}{\text{Original Value}} \times 100$$

In Ashtonville (USA), there were 450 houses. After a tornado, only 300 remained standing.

What percentage of the houses in Ashtonville was destroyed by the tornado? Give your answer to 1 decimal place.

Change in values = 450 − 300 = 150
Original number of houses = 450

Percentage change = $\frac{150}{450} \times 100 = 33.3\%$ ⬅ This is $\frac{1}{3}$.

Yesterday, Chris caught twelve fish. Today, he caught eight fewer.

What percentage decrease is this?
Give your answer to 2 decimal places.

$\frac{8}{12} \times 100 = 66.67\%$ ⬅ This is $\frac{2}{3}$.

A car park increases its daily rate from £4.00 to £4.60.

What is this increase as a percentage?

$\frac{60}{400} \times 100 = 15\%$ ⬅ Convert values to pence first.

Quick Test

1. Change 15% to **a)** a fraction and **b)** a decimal.
2. Without a calculator, work out 30% of 90 Turkish Lira.
3. Shima bought a necklace for £125. Three years later, she sold it online for £85. Calculate the percentage loss.
4. Put these values in order of size, smallest first:
   $\frac{1}{5}$, 15%, 1.15, $\frac{5}{20}$
5. James says that 20% of £60 is the same as 60% of £20. Is he correct?
6. Draw a rectangle measuring 6cm by 5cm. Divide the rectangle into 30 centimetre squares and then shade 40% of it.

Key Words

percentage

Percentages 2

You must be able to:

- Increase and decrease quantities by a percentage
- Express one quantity as a percentage of another
- Solve percentage problems using a multiplier
- Understand and use reverse percentages.

Increasing and Decreasing Quantities by a Percentage

- When **increasing** a quantity by a percentage, work out the increase and then **add** it to the original amount.
- When **decreasing** a quantity, work out the decrease and then **subtract**.
- Percentage increase and decrease problems can also be solved using a **multiplier**.

> **Key Point**
>
> Harder percentages to learn:
>
> $20\% = \frac{1}{5}$
>
> $33.3\% = \frac{1}{3}$
>
> $66.7\% = \frac{2}{3}$
>
> $12.5\% = \frac{1}{8}$

Last year, Dandelion Place had 14 240 visitors.

This year, the attraction will be closed for three months for refurbishment. This means there will be a 15% decrease in the total number of visitors for the year.

How many visitors are expected this year?

Method 1:

$\frac{15}{100} \times 14\,240 = 2136$

Number of visitors
$= 14\,240 - 2136$
$= 12\,104$

Method 2:

10% of 14 240 = 1424
 5% of 14 240 = 712 ◄──
15% of 14 240 = 2136 ◄──

so, the number of visitors
$= 14\,240 - 2136$ ◄──
$= 12\,104$

$10\% \div 2 = 5\%$

$10\% + 5\% = 15\%$

Alternatively, using a multiplier: 15% decrease means 85% of last year's visitors, so
$0.85 \times 14\,240 = 12\,104$

Molly bought a new tumble dryer.
She visited three shops to compare prices:

A. Sean's Electricals
Normal price: £250, Sale price: 5% off

B. Mandeep's Deals
Tumble dryer: £200, plus tax at 17.5%

C. Chet's Cut Price Store
Normal price: £280, Sale price: $\frac{1}{7}$ off

a) Which shop offered the best deal?
b) How much did it sell the tumble dryer for?

   A. Sean's Electricals:
   $\frac{5}{100} \times £250 = £12.50$
   Tumble dryer costs £250 − £12.50 = £237.50

B. Mandeep's Deals:
$\frac{17.5}{100} \times £200 = £35$
Tumble dryer costs £200 + £35 = £235.00

C. Chet's Cut Price Store:
$\frac{1}{7} \times £280 = £40$
Tumble dryer costs £280 – £40 = £240.00

a) Mandeep's Deals b) £235

> If working without a calculator, remember:
> 17.5% = 10% + 5% + 2.5%

Sanjeev's salary is £32000 per year. His salary is increased by 6%.

Work out his new salary.

100% + 6% = 106% = 1.06
1.06 × £32000 = £33920

> This is the multiplier.

Expressing One Quantity as a Percentage of Another

- When expressing one quantity as a percentage of another, write two quantities as a fraction and then multiply by 100 to convert to a percentage.

In two separate maths tests, Vinay got 18 out of 30 and Derek got 24 out of 45.

Who got the greater percentage in their maths test?

Vinay: $\frac{18}{30} \times 100 = 60\%$

Derek: $\frac{24}{45} \times 100 = 53.33\%$

Vinay got the greater percentage.

Work out 16 minutes as a percentage of 4 hours. Give your answer to 3 decimal places.

$\frac{16}{240}$

$\frac{16}{240} \times 100 = 6.667\%$

(to 3 d.p.)

> **Key Point**
>
> Make sure all quantities are in the same units first.

> 4 hours = 4 × 60 = 240 minutes

Reverse Percentages

- **Reverse percentages** involve working backwards from the final amount to **find the original amount**.

In a Thai restaurant, there were $27\frac{1}{2}$ dumplings left on a plate after 45% had been eaten. How many dumplings were on the plate at the start of the meal?

100% – 45% = 55%
1% = 27.5 ÷ 55
100% = 100 × 27.5 ÷ 55 = 50 dumplings

> $27\frac{1}{2}$ dumplings is 55% of the original quantity.

Quick Test

1. This year Curt grew 220 carrots. This is 20% less than last year. How many carrots did Curt grow last year?
2. Increase £60 by 17.5%.
3. Express 18cm as a percentage of 12m.

> **Key Words**
>
> multiplier

Probability 1

You must be able to:

- Know and use words associated with probability
- Construct and use a probability scale
- Understand what mutually exclusive events are
- Calculate probabilities using experimental data.

Calculating Probabilities

- The **probability** of an outcome occurring can be described using words or using a numerical scale from 0 to 1.

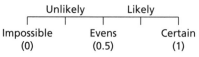

- **Relative frequencies** are probabilities based on experiments.
- **Random** means each possible outcome is equally likely.
- An event is **biased** when outcomes are **not** equally likely.
- The sample space represents all possible outcomes from an event. This can be shown as a list or a diagram.

> There are seven counters in a bag.
> Four counters are red, two are green and one is blue.
>
> One counter is taken from the bag at random.
>
> Write down the probability that the counter taken is:
>
> a) Red   Four of the counters are red, so P(red) = $\frac{4}{7}$
>
> b) Green Two of the counters are green, so P(green) = $\frac{2}{7}$
>
> c) Blue.   One of the counters is blue, so P(blue) = $\frac{1}{7}$

> **Key Point**
>
> Probability is the chance that an event is likely to occur.
>
> Probabilities can be based on theory or the results of an experiment.
>
> The sum of the probabilities of all possible outcomes is 1.

> **Key Point**
>
> Probabilities can be given as fractions, decimals or percentages.

> There are seven counters, so seven possible outcomes. Each outcome is equally likely.

> Note: $\frac{4}{7} + \frac{2}{7} + \frac{1}{7} = 1$

Mutually Exclusive and Exhaustive Outcomes

- **Mutually exclusive** outcomes **cannot** happen at the same time.
- When two events are mutually exclusive P(A or B) = P(A) + P(B).
- A list of **exhaustive** events contains all possible outcomes.
- The probabilities of a set of exhaustive outcomes add up to 1.

> Shelby looks at the weather forecast every morning.
> On Monday, the forecast says that there is a $\frac{7}{10}$ chance of rain.
>
> What is the probability that it will **not** rain?
>
> P(no rain) = $1 - \frac{7}{10} = \frac{3}{10}$

> **Key Point**
>
> P(A') is the probability of A **not** happening
>
> P(A') = 1 – P(A)

A spinner has segments that are coloured red, blue, green or yellow. The probability that it will land on a particular colour is shown below:

Colour	Red	Blue	Green	Yellow
Probability	0.4	x	0.2	0.3

What is the probability that the spinner will land on blue?

$$0.4 + x + 0.2 + 0.3 = 1$$
$$x = 1 - 0.9$$
$$= 0.1$$
$$P(\text{blue}) = 0.1$$

Expectation

- When you know the probability of an outcome, you can predict how many times you would expect that outcome to occur in a certain number of trials. This is called **expectation**.

Kelly rolls a six-sided dice 120 times and records her results.

Score	1	2	3	4	5	6
Frequency	19	29	14	18	20	20

a) Kelly throws the dice again.

Estimate the probability that she will throw a 2.

Estimate of probability of a 2 is $\frac{29}{120}$.

b) The dice is thrown another 200 times.

How many times would you expect it to land on a 2?

$\frac{29}{120} \times 200 = 48$

c) Is there enough evidence to suggest the dice is biased? Explain your answer.

If the dice is fair, you would expect to roll 20 of each number. This did not happen, so the dice may be biased.

Quick Test

1. A spinner has four faces labelled 1, 2, 3 and 4. The probability that it will land on a 2 or 4 is given in the table.

Number	1	2	3	4
Probability	x	0.3	x	0.5

a) Work out the value of x.
b) If you spin the spinner 150 times, how many times would you expect it to land on a 2?
c) Is the spinner biased? Give a reason for your answer.

Probability 2

You must be able to:

- Calculate the probability of independent and dependent combined events
- Calculate probabilities using sample space and tree diagrams
- Calculate sets and combinations of sets using tables, grids and Venn diagrams.

Probability Diagrams

Julie thinks that in her class, you are more likely to wear glasses if you are a boy. Based on the data below, is she correct?

	Boys	Girls	Total
Glasses	8	6	14
No Glasses	12	9	21
Total	20	15	35

P(of a boy wearing glasses) = $\frac{8}{20} = \frac{2}{5}$

P(of a girl wearing glasses) = $\frac{6}{15} = \frac{2}{5}$

Julie is incorrect as the probabilities are equal.

Cancel down fractions.

> **Key Point**
>
> To calculate the probabilities involving more than one event, you can use a **two-way table**, **Venn diagram** or tree diagram to represent all the possible outcomes.

Students in a school study either French, Spanish, both or neither. There are 250 students. 75 students study both French and Spanish, 225 study French and 90 study Spanish.

Work out how many students do **not** study French or Spanish.

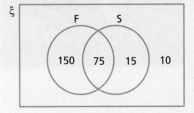

$250 - 150 - 75 - 15 = 10$

10 students do not study French or Spanish.

The crossover represents students who study both subjects.

Combined Events

- When calculating probabilities for combined events, you can draw a **sample space diagram**.

Two fair six-sided dice are thrown and their scores are added together.

a) Construct a sample space diagram to show all the possible outcomes.

	Dice 1					
+	**1**	**2**	**3**	**4**	**5**	**6**
1	2	3	4	5	6	7
2	3	4	5	6	7	8
3	4	5	6	7	8	9
4	5	6	7	8	9	10
5	6	7	8	9	10	11
6	7	8	9	10	11	12

(Dice 2 labels the rows)

b) Using your sample space diagram, work out P(8).

The sample space diagram shows there are 6 × 6 = 36 possible outcomes.

8 appears five times in the diagram out of 36 possible outcomes, therefore $P(8) = \frac{5}{36}$

Tree Diagrams

- When dealing with successive events you can use a **tree diagram**.

Two fair coins are tossed one after the other.

Write down the probability of each possible outcome.

```
              0.5      H  P(HH) = 0.5 × 0.5 = 0.25
      0.5    ╱
      ╱   H ─────────  T  P(HT) = 0.5 × 0.5 = 0.25
     ╱       0.5
    ╱         0.5      H  P(TH) = 0.5 × 0.5 = 0.25
     ╲       ╱
      ╲   T ─────────
      0.5    0.5      T  P(TT) = 0.5 × 0.5 = 0.25
```

Each coin can give us two possible outcomes: a Head (H) or a Tail (T).

Probability of two Heads = P(HH) = 0.25
Probability of two Tails = P(TT) = 0.25
Probability of a Head and a Tail = P(HT) + P(TH)
$$= 0.25 + 0.25$$
$$= 0.5$$

0.25 + 0.25 + 0.5 = 1

> **Key Point**
>
> Always multiply along the branches of a probability tree.

Independent Events

- If A and B are **independent** events, one event does **not** depend on the other.
- If two events A and B are independent then P(A and B) = P(A) × P(B).

A fair two-sided coin is tossed and a fair six-sided dice is rolled.

What is the probability of getting a Head and rolling a 6?

$$P(H) \times P(6) = \frac{1}{2} \times \frac{1}{6}$$
$$= \frac{1}{12}$$

The two events are independent, therefore: P(A) × P(B)

A fair six-sided dice is rolled and a card is taken at random from a pack of 52 playing cards.

What is the probability of getting a 5 and a king?

$$P(5) \times P(\text{king}) = \frac{1}{6} \times \frac{4}{52} = \frac{4}{312} = \frac{1}{78}$$

> **Quick Test**
>
> 1. A fair three-sided spinner has one red section, one green section and one blue section all of equal size. The spinner is spun and a fair six-sided dice is thrown. What is the probability of getting a 5 on the dice with a red on the spinner?
> 2. A bag contains three red balls, four blue balls and two yellow balls. A ball is drawn from the bag and then replaced. A second ball is then drawn from the bag.
>   What is the probability that both balls are the same colour?

> **Key Words**
>
> two-way table
> Venn diagram
> sample space diagram
> tree diagram
> independent

Number 1, 2 & 3

1 The average distance from the Moon to the Earth is 384 000km.

   a) Write this distance in standard form. [2]

   b) A spaceship is travelling from the Earth to the Moon.
      It has travelled 1.96×10^4km.

      How many kilometres are left to travel? [2]

2 On one evening in December, the temperature in Birmingham was $-3°C$ and the
   temperature in Aberdeen was $-8°C$.

   Work out the difference in temperature between Aberdeen and Birmingham. [1]

3 Which is larger?
   Circle your answer.

   $3 + 6 \times 2$ OR $(-3 - 6) - (-25)$ [2]

4 What is the largest multiple of 4 and 7 that is smaller than 105? [1]

5 What is the lowest common multiple (LCM) of 12 and 15?
   Circle the correct number.

   3   180   60   1 [1]

6 Which is greater, the fifth square number OR the third cube number? [3]

7 256 can be expressed as 2^n.

   What is the value of n? [2]

8 Is 2^3 equal to 3^2?
   Explain your answer. [2]

9 Sausages come in packs of six. Bread rolls come in packs of four.

   What is the least number of packs of sausages and packs of bread rolls that Linda must buy,
   so that there is a roll for every sausage without any left over? [2]

Total Marks _____ / 18

Basic Algebra & Factorisation and Formulae

1 Expand $5(x + 6)$ [2]

2 Factorise $15x + 10$ [1]

3 Expand $6(5 - 2x)$ [2]

4 Make t the subject of $p = 4t - q$ [2]

5 Simplify $5x - 2y + 4x + 6y$ [2]

6 Simplify $5 - 3z + y - 5z + 7 + 3y$ [2]

7 Simplify $3x^2 + 3x + x^2 + 4 - x$

Circle your answer.

$10x + 4$ $\qquad$ $4x^2 + 2x + 4$ $\qquad$ $6x^2 + 4$ $\qquad$ $10x^2$ [1]

8 Work out the value of $3z^2 - 2q + 5$, when $z = -2$ and $q = -3$.

Circle your answer.

11 $\qquad$ 23 $\qquad$ 35 $\qquad$ 47 [1]

9 Solve $4(2b - 3) = 2$ [2]

10 Solve $3(p + 2) = 2(p + 3)$ [2]

11 Solve $\frac{5}{2}x - \frac{1}{3} = \frac{2}{3}x + \frac{1}{2}$ [3]

12 Expand $6(x - 5y + 6)$ [2]

13 Expand $6p - 4(q - 3)$ [2]

14 Factorise completely $4xyz - 4xz$ [2]

15 Factorise $x^2 + 3x + 2$ [2]

16 Write $3(2x - 5) + 4(x + 3) - 4x$ in the form $a(bx + c)$, where a, b and c are integers. [3]

17 Look at the options below and circle the formula.

$x^4 - x^2$ $\qquad$ $2x - x = x$ $\qquad$ $3x - 5 = 10$ $\qquad$ $A = \frac{1}{2}(a + b)h$ [1]

Total Marks ...... / 32

Practice Questions

Ratio and Proportion

1 Simplify 5g : 10kg 🖩 [1]

2 The angles in a triangle are in the ratio 2 : 3 : 4

What is the size of the largest angle? 🖩 [2]

3 Six sticks of celery are needed to make celery soup for four people.

How many sticks of celery would be needed to make soup for 20 people? 🖩 [2]

4 It took six people four days to build a wall.

Working at the same rate, how long would it have taken one person to build the wall? [2]

> **Total Marks** _____ / 7

Variation and Compound Measures

1 A bar of lead has a volume of 400cm³ and a mass of 4560g.

Work out the density of the bar of lead in g/cm³. [2]

2 A rabbit runs 200 metres in 22 seconds.

What is the rabbit's average speed in m/s?
Give your answer to 2 decimal places. [2]

3 Khalid left his home at 10am and went for a 15km run.
He arrived back home at 1pm.

What was his average speed in km/h? [2]

4 Work out the compound interest earned on £4000 invested at 4% for four years. [3]

> **Total Marks** _____ / 9

Angles and Shapes 1 & 2

1 Work out the size of angles *j*, *k*, *l* and *m*, giving a reason for each answer.

[4]

2 Three angles in a quadrilateral are 46°, 107° and 119°.

Calculate the size of the fourth angle. [1]

3 Work out the exterior angle of a regular decagon. [1]

4 *A* and *B* are two points.

If the bearing of *B* from *A* is 036°, what is the bearing of *A* from *B*? [1]

5 A map is drawn using a scale of 1cm : 4km.

If the length of Loch Ness is 36km, what would its length be on the map? [1]

6 A boat sails in a north-westerly direction.

What bearing is this? [1]

Total Marks _____ / 9

Practice Questions

Fractions

1 A teacher took 32 books home to mark. She marked $\frac{1}{8}$ of them.

How many books does she still have to mark? [2]

2 On a farm, $\frac{1}{3}$ of the livestock is cows, $\frac{1}{6}$ is sheep, $\frac{1}{4}$ is chickens and the remainder is horses.

Express how many horses there are on the farm as a fraction. [2]

3 Write these fractions in order, smallest first:

$\frac{2}{3}$ $\frac{4}{12}$ $\frac{1}{6}$ $\frac{12}{24}$ [2]

4 **a)** Work out the area of a rectangle measuring $\frac{4}{5}$ of a metre by $\frac{2}{3}$ of a metre. [1]

     b) What is the rectangle's perimeter? Give your answer as an improper fraction. [2]

5 Write down:

     a) 0.45 as a fraction in its simplest form. [2]

     b) $\frac{7}{8}$ as a decimal. [2]

Total Marks _____ / 13

Percentages 1 & 2

1 Whizzy Garage sells cars. It offers a discount of 20% off the retail price for cash purchases. Pat pays £4800 cash for a car.

Calculate the retail price of the car before the cash discount. [2]

2 In a box containing 36 bananas, nine of the bananas are bad.

What percentage of the bananas are bad? [2]

3 What is 28 minutes of 2 hours as a percentage?
Circle your answer.

2.61%   7.14%   14%   23.33% [1]

Total Marks _____ / 5

Probability 1 & 2

1   Two fair six-sided dice are thrown and the product of their scores is calculated. 🖩

    a) Draw a sample space diagram to represent all the possible outcomes. [2]

    b) What is the probability that the product of the scores is greater than 12? [1]

    c) What is the probability that the product of the scores is a square number? [1]

2   There are 20 pens in a box: ten black, six blue and four red.
    A pen is picked at random from the box. 🖩

    a) What is the probability that the pen is black? [1]

    b) What is the probability that the pen is **not** black? [1]

3   A three-sided spinner has sides labelled 1, 2 and 3. The spinner is biased.
    Thomas spins the spinner 500 times
    and records the outcomes in a table.

Score	1	2	3
Frequency	98	305	97

    a) Thomas spins the spinner again. Estimate the probability that it lands on a 2. [1]

    b) He takes another 100 spins. Estimate how many times it will land on a 3. [2]

4   Ten coloured counters are placed in a bag.
    Six counters are pink, three counters are green and one counter is blue. 🖩

    a) One counter is taken from the bag at random.

        Work out the probability that the counter taken is:
        i) Pink   ii) Green   iii) Blue. [3]

    b) A counter is taken from the bag at random and replaced.
        A second counter is then taken from the bag at random.

        Work out the probability that both counters are the same colour. [2]

5   Bhavna conducts a survey about the pets owned by her classmates.
    There are 35 students in her class. Her results are: 25 students own cats,
    15 students own dogs and 10 students own both. 🖩

    a) Draw a Venn diagram to represent the results. [3]

    b) How many students do **not** own a cat or a dog? [1]

Total Marks _____ / 18

Number Patterns and Sequences 1

You must be able to:

- Work out missing terms in sequences using term-to-term rules and position-to-term rules
- Recognise and use arithmetic and geometric sequences
- Work out the rule for a given pattern.

Patterns in Number

- A **sequence** is a series of shapes or numbers that follow a particular pattern or rule.
- A **term-to-term rule** links the next term in the sequence to the previous one.
- A **position-to-term rule**, also called the *n*th term, can be used to work out any term in the sequence.

> Write down the next two terms in the following sequence:
>
> 7, 11, 15, 19, __, __
>
> The term-to-term rule is +4, so the next two terms are 23 and 27.

> Write down the next two terms in the following sequence:
>
> 50, 25, 12.5, __, __
>
> The term-to-term rule is ÷2, so the next two terms are 6.25 and 3.125.

General Rules from Given Patterns

> Here is a sequence of patterns made from matchsticks:
>
>
>
> pattern 1   pattern 2   pattern 3
>
> **a)** Draw the next pattern in the sequence.
>
>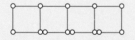
>
> **b)** Write down the sequence of numbers that represents the total number of matchsticks used in each pattern and state the term-to-term rule.
>
> | Pattern No. | 1 | 2 | 3 | 4 |
> |---|---|---|---|---|
> | No. of Matchsticks | 4 | 7 | 10 | 13 |
>
> The term-to-term rule is +3.

> **Key Point**
>
> When finding the term-to-term rule, remember to look at consecutive terms.

The nth term of a number sequence is $5n + 2$.

Write down the first five terms of the sequence.

$(5 \times 1) + 2 = 7$ ← To work out the first term, substitute $n = 1$ into the expression.

$(5 \times 2) + 2 = 12$ ← To work out the second term, substitute $n = 2$ into the expression.

7, 12, 17, 22, 27 ← Continue until $n = 5$ to produce the first five terms of the sequence.

Number Sequences

- In an **arithmetic sequence**, there is a common difference between consecutive terms, e.g.

  5, 8, 11, 14, 17 … ← The terms have a common difference of +3.

- In a **geometric sequence**, each term is found by multiplying the previous term by a constant, e.g.

  20, 10, 5, 2.5, 1.25 … ← The constant (or ratio) is 0.5

The nth term of an arithmetic sequence is $4n - 1$.

a) Write down the term-to-term rule.

The sequence of numbers is 3, 7, 11, 15, 19 … ← Work out the first five terms.
The term-to-term rule is +4.

b) Marnie thinks that 50 is a number in this sequence.
Is Marnie correct? Give a reason for your answer.

$4n - 1 = 50$, $n = 12.75$ ← Use the nth term to find the value of n for an output of 50.
n is not a whole number, so 50 is **not** in this sequence.
Marnie is wrong.

Here is a geometric sequence: 4, 6, 9, __, 20.25

What is the missing term?

$\frac{6}{4} = \frac{9}{6} = 1.5$ ← Divide at least two given terms by the previous term to work out the ratio.

$9 \times 1.5 = 13.5$ ← Multiply by the ratio to find the missing term.

The missing term is 13.5

Quick Test

1. Here are the first five terms of a sequence: 16, 12, 8, 4, 0 …
   Write down the next two terms.
2. The nth term of a sequence is $5n - 7$.
   Work out the 1st term and the 10th term of this sequence.
3. In the sequence below the next pattern is formed by adding another layer of tiles around the previous pattern:

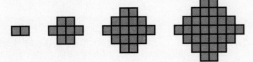

   Work out how many tiles will be needed for the 6th pattern.

Key Words

sequence
term-to-term rule
position-to-term rule
nth term
arithmetic sequence
geometric sequence

Number Patterns and Sequences 2

You must be able to:

- Work out and use expressions to calculate the nth term of a linear sequence
- Recognise and use sequences of triangular numbers, squares, cubes and other special sequences.

Finding the nth Term of a Linear Sequence

- A number sequence that increases or decreases by the same amount each time is called a **linear sequence**.
- To work out the expression for the nth term in a linear sequence, look for a pattern in the numbers.
- Using a function machine to represent a sequence of numbers can help.

The first five terms of a sequence are: 9, 12, 15, 18, 21 …

What is the expression for the nth term of this sequence?

Input (n)	× 3 ($3n$)	Output
1	3	9
2	6	12
3	9	15
4	12	18
5	15	21
n	$3n$	$3n + 6$

The difference between $3n$ and the output in each case is 6, so the expression for the nth term is **$3n + 6$.**

The 'input' is the position of the term and the 'output' is the value of the term.

The term-to-term rule is +3, so the expression for the nth term starts with $3n$.

- The alternative method is to work out the **zero term**.

The first five terms of a sequence are: 20, 16, 12, 8, 4 …

What is the expression for the nth term of this sequence?

Input	Output
0	zero term
1	20
2	16
3	12
4	8
5	4
n	nth term

The expression for the nth term is $-4n + 24$ or $24 - 4n$.

> ### Key Point
>
> The zero term is the term that would come before the first term in a given sequence of numbers.

The zero term is (20 + 4 =) 24

The difference between terms is –4.

The nth term = 'the difference' × n + the zero term

Special Sequences

- It is important to be able to recognise special sequences of numbers, e.g. the square numbers, the cube numbers, the triangular numbers and the **Fibonacci sequence**.

Here is a sequence of numbers: 1, 4, 9, 16, 25 …

Write down:

a) The next two terms.

36, 49 ←——————————————————— The sequence is the square numbers.

b) The nth term of the sequence.

The nth term is n^2.

Here is a sequence of numbers: 1, 8, 27, 64 …

Write down:

a) The next two terms.

125, 216 ←——————————————————— The sequence is the cube numbers.

b) The nth term of the sequence.

The nth term is n^3.

Here is a sequence of numbers: 1, 3, 6, 10, 15 …

Write down:

a) The next two terms.

21, 28 ←——————————————————— The sequence is the triangular numbers.

b) The nth term of the sequence.

The nth term is $\frac{n}{2}(n+1)$.

Write down the next two terms in the following sequence:
1, 1, 2, 3, 5, 8, 13, 21 …

34, 55 ←——————————————————— This is the Fibonacci sequence. The next term is found by adding the previous two terms together.

> **Quick Test**
>
> 1. Write down the next two terms in the following sequence:
>   6, 9, 12, 15, __, __
> 2. a) Write down the nth term for the following sequence:
>   7, 10, 13, 16, 19 …
>   b) Work out the 50th term in this sequence.
> 3. Write down the nth term for the following sequence:
>   0, 3, 8, 15, 24 …

> **Key Words**
>
> linear sequence
> zero term
> Fibonacci sequence

Transformations

You must be able to:

- Identify, describe and construct transformations of shapes, including reflections, rotations, translations and enlargements.

Reflection

- When a shape is reflected:
  - Each point on the image is the same distance from the mirror line as the corresponding point on the object
  - The object and the image are **congruent** (same size and shape)
  - To define a **reflection** on a coordinate grid, the equation of the mirror line should be stated.

Rotation

- **Rotation** is described by stating the:
  - Direction rotated (clockwise or anticlockwise)
  - Angle of rotation (which is usually a multiple of 90° in the exam)
  - Centre of rotation (point about which the shape is rotated).

> **Key Point**
>
> There is no need to state clockwise or anticlockwise for a rotation of 180°.

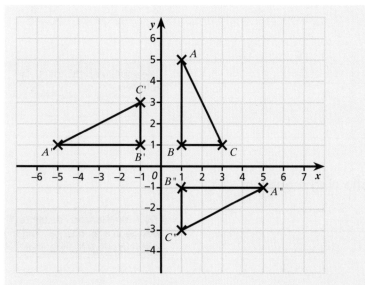

a) Describe the **transformation** that maps triangle ABC onto triangle $A'B'C'$.

The transformation that maps ABC to $A'B'C'$ is a rotation 90° anticlockwise (or 270° clockwise) about the origin (0, 0).

b) Describe the transformation that maps triangle $A'B'C'$ onto triangle $A''B''C''$.

The transformation that maps $A'B'C'$ to $A''B''C''$ is a rotation of 180° about the origin (0, 0).

Translation

- When a shape is translated:
  - The shape does not rotate – it moves left or right and up or down – and stays the same size
  - The **translation** is represented by a **column vector** $\begin{pmatrix} x \\ y \end{pmatrix}$.

Describe the transformation that takes shape A to shape B.

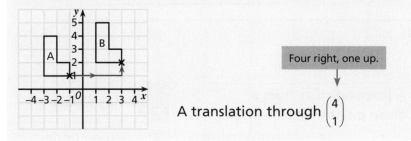

Four right, one up.

A translation through $\begin{pmatrix} 4 \\ 1 \end{pmatrix}$

Enlargement

- When a shape is enlarged:
  - The shape of the object is not changed, only its size
  - The enlarged shape is **similar** to the original shape
  - The **scale factor** determines whether the object gets bigger (scale factor > 1) or smaller (scale factor < 1)
  - When describing the **enlargement**, state the scale factor and the centre of enlargement.

a) Enlarge triangle A by scale factor 2, centre of enlargement (1, 2). Label the transformed triangle, B.

b) Enlarge triangle A by scale factor $\frac{1}{2}$, centre of enlargement (1, 2). Label the transformed triangle, C.

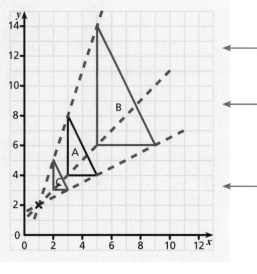

All construction lines must remain.

The side **lengths** of triangle B are **twice** the length of the corresponding sides of triangle A. However, the **area** of triangle B is **four times** bigger.

The side **lengths** of triangle C are **half** the length of the corresponding sides of triangle A. However, the **area** of triangle C is **four times** smaller.

Quick Test

1. Describe the single transformation that takes:
    a) Triangle A to triangle B
    b) Triangle A to triangle C
    c) Triangle A to triangle D.

Constructions

You must be able to:

- Use a ruler and a pair of compasses to produce different constructions, including bisectors
- Describe a locus and solve problems involving loci.

Constructions

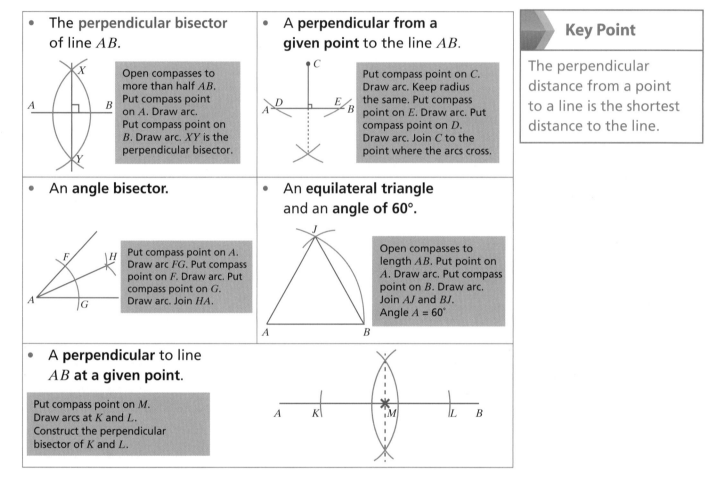

- The **perpendicular bisector** of line AB.

  Open compasses to more than half AB. Put compass point on A. Draw arc. Put compass point on B. Draw arc. XY is the perpendicular bisector.

- A **perpendicular from a given point** to the line AB.

  Put compass point on C. Draw arc. Keep radius the same. Put compass point on E. Draw arc. Put compass point on D. Draw arc. Join C to the point where the arcs cross.

Key Point

The perpendicular distance from a point to a line is the shortest distance to the line.

- An **angle bisector**.

  Put compass point on A. Draw arc FG. Put compass point on F. Draw arc. Put compass point on G. Draw arc. Join HA.

- An **equilateral triangle** and an **angle of 60°**.

  Open compasses to length AB. Put point on A. Draw arc. Put compass point on B. Draw arc. Join AJ and BJ. Angle $A = 60°$

- A **perpendicular** to line AB **at a given point**.

  Put compass point on M. Draw arcs at K and L. Construct the perpendicular bisector of K and L.

Defining a Locus

- A **locus** is the path taken by a point that is obeying certain rules.
- The plural of locus is **loci**.

Key Point

In the exam, the fixed distance is likely to be given, e.g. 4cm.

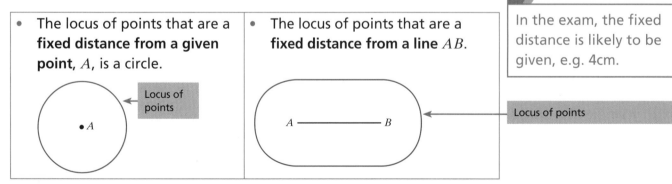

- The locus of points that are a **fixed distance from a given point**, A, is a circle.

  Locus of points

- The locus of points that are a **fixed distance from a line** AB.

  Locus of points

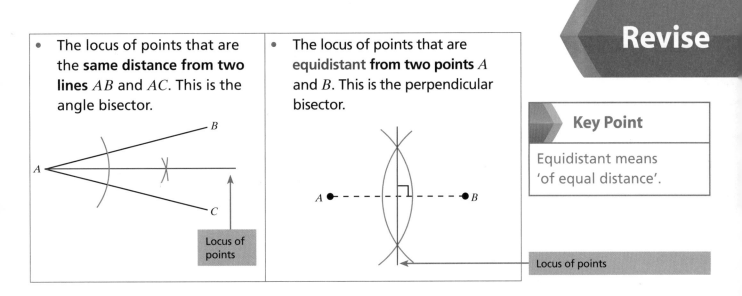

- The locus of points that are the **same distance from two lines** AB and AC. This is the angle bisector.

- The locus of points that are **equidistant from two points** A and B. This is the perpendicular bisector.

Locus of points

Locus of points

Loci Problems

A guard dog is tied to a post by a 4-metre long rope.

Accurately draw the locus of the points the dog can reach using a scale of 1cm : 1m.

The solution would be a shaded circle of radius 4cm. The dog could reach the circumference of the circle and any point inside it.

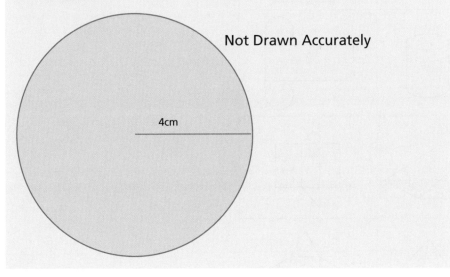

Not Drawn Accurately

4cm

Quick Test

1. Describe how you would accurately construct an angle of 30°.
2. A rectangular vegetable plot $ABCD$ measures 6m by 3m. A goat is tethered at corner A by a 4m rope. Accurately draw the plot and construct the locus of the points that the goat can reach. Scale 1cm : 1m. Shade the region of the vegetable plot that can be eaten by the goat.

Key Words

perpendicular
bisector
locus / loci
equidistant

Nets, Plans and Elevations

You must be able to:

- Identify a 3D shape from its net
- Draw nets of 3D shapes
- Use isometric grids
- Interpret and draw plans and elevations of 3D shapes.

Nets

- A net is a 2D shape that can be folded to form a 3D solid.

Name	Shape	Net
Cube		
Cuboid		
Cylinder		
Triangular Prism		
Square-Based Pyramid		

> **Key Point**
>
> There are many different nets for a cube.

Plans and Elevations

- An **isometric grid** can be used to draw 3D shapes.
- The **plan view** of a 3D shape shows what it looks like from above, i.e. a bird's eye view.
- The side **elevation** is the view of a 3D shape from the side.
- The front elevation is the view of a 3D shape from the front.

> **Key Point**
>
> An isometric grid can be made up of equilateral triangles or dots.

Here is a 3D shape made of centimetre cubes:

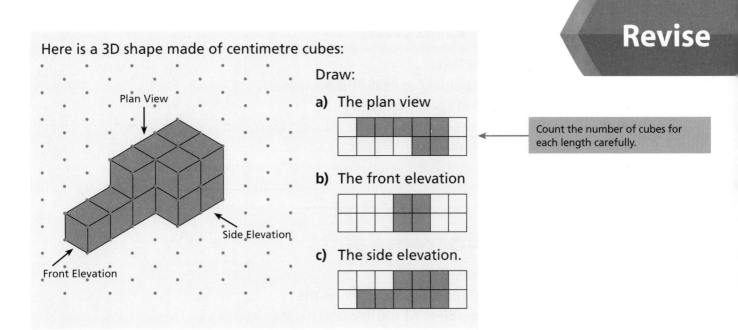

Plan View

Front Elevation

Side Elevation

Draw:

a) The plan view

Count the number of cubes for each length carefully.

b) The front elevation

c) The side elevation.

Here are the three views of a 3D shape:

Plan View

Front Elevation

Side Elevation

Draw the 3D shape on isometric paper.

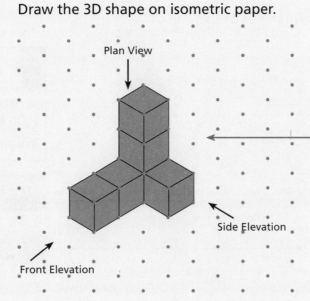

Plan View

Side Elevation

Front Elevation

Check that your final drawing matches the given views.

Quick Test

1. A can of soup has a height of 10cm. The length of the label required to go once around the can is 22cm.

Draw an accurate plan view and front elevation of the can of soup.

Key Words

net
isometric grid
plan view
elevation

Linear Graphs

You must be able to:

- Work with coordinates in all four quadrants
- Plot graphs of linear functions
- Work out the equation of a line through two given points or through one point with a given gradient
- Work out the gradient and y-intercept of a straight line in the form $y = mx + c$.

Drawing Linear Graphs from Points

- **Linear graphs** are straight-line graphs.
- The equation of a straight-line graph is usually given in the form $y = mx + c$, where m is the **gradient** of the line and c is the **intercept** of the y-axis.
- $y = mx + c$ is a function of x, where the **input** is the x-coordinate and the **output** is the y-coordinate.

> Write down the gradient and y-intercept of the line of equation $y = 5x + 1$.
>
> The gradient is 5.
> The y-intercept is (0,1).

Key Point

To draw a straight line, only two coordinates are needed.

> Draw the graph of the equation $y = 2x + 5$.
> Use values of x from –3 to 3.

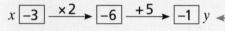

You can set up a flow chart to work out the y-coordinates.

x	–3	0	3
y	–1	5	11

Draw a table of values. Include a third value as a check.

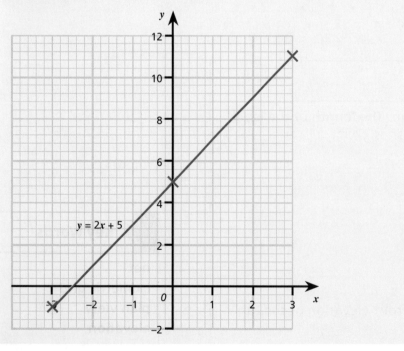

The Cover-Up Method

- Another method that can be used to draw a graph is the cover-up method.
- This can be used for equations in the form $ax + by = c$.

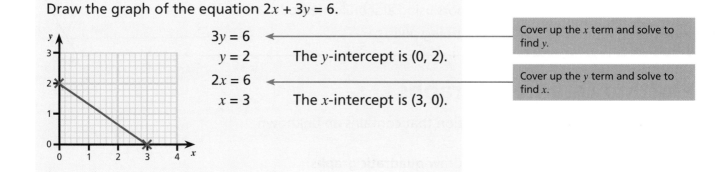

Draw the graph of the equation $2x + 3y = 6$.

$$3y = 6$$
$$y = 2 \qquad \text{The } y\text{-intercept is } (0, 2).$$
$$2x = 6$$
$$x = 3 \qquad \text{The } x\text{-intercept is } (3, 0).$$

> Cover up the x term and solve to find y.

> Cover up the y term and solve to find x.

Finding the Equation of a Line

- To find the equation of a straight line in the form $y = mx + c$, work out the gradient and y-intercept.

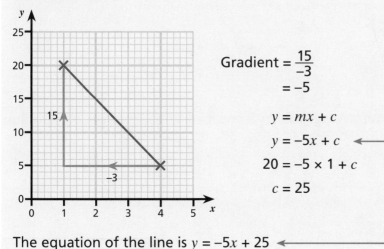

Work out the equation of the line that joins the points $(1, 20)$ and $(4, 5)$.

$$\text{Gradient} = \frac{15}{-3}$$
$$= -5$$

$$y = mx + c$$
$$y = -5x + c$$
$$20 = -5 \times 1 + c$$
$$c = 25$$

> To work out the value of c, substitute in the gradient and point $(1, 20)$ or $(4, 5)$.

The equation of the line is $y = -5x + 25$

> Substitute your values for m and c into the equation.

Key Point

$$\text{Gradient} = \frac{\text{Change in } y}{\text{Change in } x}$$

If the line slopes down from left to right, the gradient is negative.

See p.76–77 for more information on gradients.

Quick Test

1. Draw the graph with equation $y = 4x + 1$ from $x = 0$ to $x = 6$.
2. Write down the gradient and y-intercept of the line with equation $y = 5 - 2x$.
3. Work out the equation of the line that joins the points $(5, 7)$ and $(3, 10)$.

Key Words

linear graph
gradient
intercept
input
output

Graphs of Quadratic Functions

You must be able to:

- Recognise, sketch and interpret graphs of quadratic functions
- Identify and interpret roots, intercepts and turning points of quadratic functions
- Work out roots using algebraic methods
- Work out turning points.

Plotting Quadratic Graphs

- A **quadratic equation** is an equation that contains an unknown term with a power of 2, e.g. x^2.
- You can use a table of values to draw **quadratic graphs**.

Draw the graph of the function $y = 2x^2 + 1$.

$x \boxed{-2} \xrightarrow{x^2} \boxed{4} \xrightarrow{\times 2} \boxed{8} \xrightarrow{+1} \boxed{9} \, y$

x	−2	−1	0	1	2	3	4
y	9	3	1	3	9	19	33

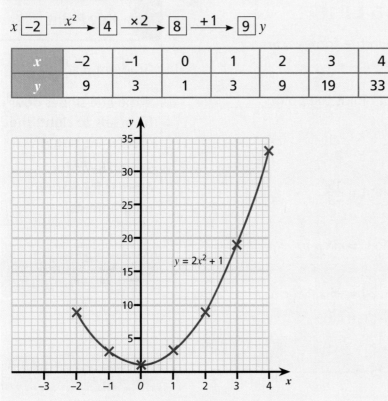

$y = 2x^2 + 1$

The Significant Points of a Quadratic Curve

- A sketch is used to show the shape and significant points on a graph, but it is not an accurate drawing.
- To sketch a quadratic, work out the **roots**, the **intercept** and the **turning point**, i.e. **maximum** or **minimum point**.
- The roots are found by solving the quadratic when $y = 0$.
- Because quadratic curves are symmetrical, the turning point is halfway between the two roots.

Sketch the graph of equation $y = x^2 + 5x + 4$.

Roots:

$x^2 + 5x + 4 = 0$

Work out the values for x when $y = 0$.

$(x + 4)(x + 1) = 0$

$x = -4$ and $x = -1$

y-intercept:

$y = 4$

y-intercept is $(0, 4)$.

Substitute $x = 0$ into the equation.

Turning Point:

$x = -4$ and $x = -1$, therefore $x = -2.5$

The value of x for the turning point is in the middle of the two roots.

$y = (-2.5)^2 + (5 \times -2.5) + 4$

$\quad = -2.25$

Turning point is $(-2.5, -2.25)$.

Substitute the value of x into the equation.

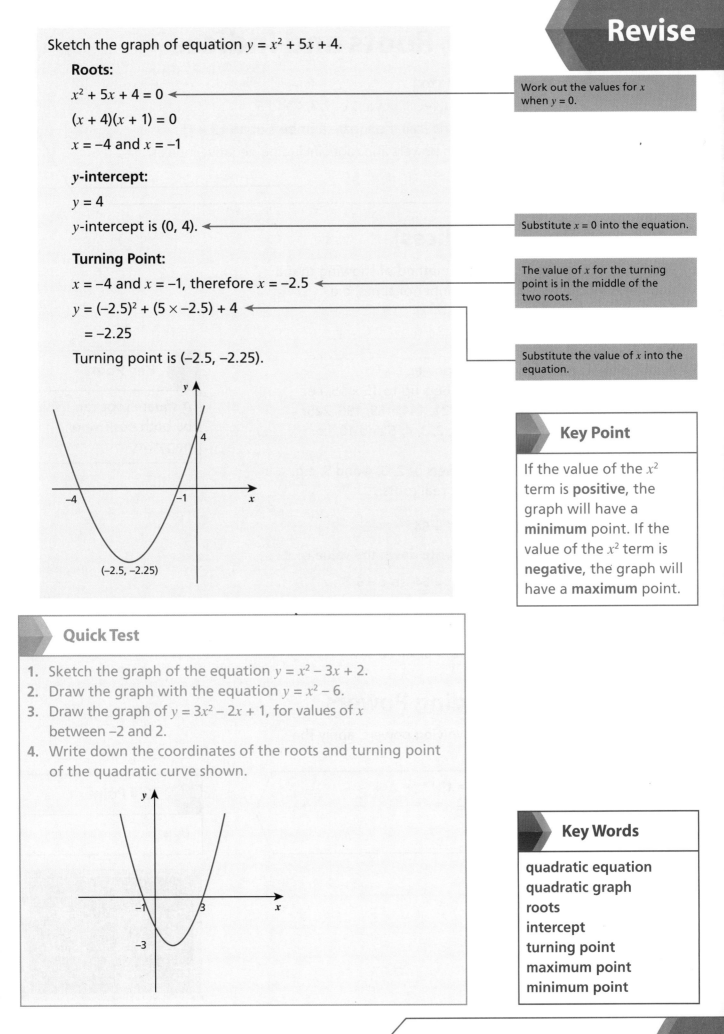

(−2.5, −2.25)

Key Point

If the value of the x^2 term is **positive**, the graph will have a **minimum** point. If the value of the x^2 term is **negative**, the graph will have a **maximum** point.

Quick Test

1. Sketch the graph of the equation $y = x^2 - 3x + 2$.
2. Draw the graph with the equation $y = x^2 - 6$.
3. Draw the graph of $y = 3x^2 - 2x + 1$, for values of x between −2 and 2.
4. Write down the coordinates of the roots and turning point of the quadratic curve shown.

Key Words

quadratic equation
quadratic graph
roots
intercept
turning point
maximum point
minimum point

Powers, Roots and Indices

You must be able to:

- Recognise and recall powers of 2, 3, 4 and 5
- Recognise and recall the square numbers up to 15 × 15
- Calculate with powers and roots, including negative indices.

Roots and Powers (Indices)

- Powers or **indices** are a shorthand method of showing that a number is multiplied by itself a number of times, e.g.
  $5 \times 5 = 5^2$
  $5 \times 5 \times 5 = 5^3$
  $5 \times 5 \times 5 \times 5 = 5^4$, etc.
- A **root** is the inverse function of a power.
- You must learn all the square numbers up to 15 × 15, i.e. 1, 4, 9, 16, 25, 36, 49, 64, 81, 100, 121, 144, 169, 196, 225.
- You must also learn the cubes of 1, 2, 3, 4, 5 and 10, i.e. 1, 8, 27, 64, 125 and 1000.
- You must be able to recognise powers of 2, 3, 4 and 5, e.g. $16 = 2^4$ and $243 = 3^5$, and work out real roots.

> **Key Point**
>
> A square root can be both positive or negative.

Write down the value of 10^6.	$2^x = 64$
$10^6 = 10 \times 10 \times 10 \times 10 \times 10 \times 10$ $\quad = 1\,000\,000$	Write down the value of x. $2^6 = 64$, so $x = 6$
Work out the value of 3^4.	Write down the value of $\sqrt[3]{8}$.
$3 \times 3 \times 3 \times 3 = 81$	$2^3 = 8$, so $\sqrt[3]{8} = 2$

Multiplying and Dividing Powers

- When completing calculations involving powers, apply the following rules:

$$x^m \times x^n = x^{m+n}$$
$$x^m \div x^n = x^{m-n}$$
$$(x^m)^n = x^{mn}$$

LEARN

> **Key Point**
>
> $x^0 = 1$

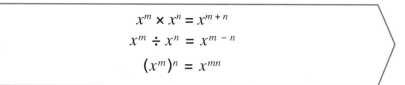

$(a^m)^n = a^{mn}$

Power of a Power

Write each of the following as a single power of 2.

a) $2^3 \times 2^4$

$= 2^{3+4} = 2^7$

b) $2^9 \div 2^4$

$= 2^{9-4} = 2^5$

c) $(2^3)^4$

$= 2^{3 \times 4} = 2^{12}$

d) $2^3 \times 2^2 \times 2^4$

$= 2^{3+2+4} = 2^9$

e) Write the following as a single power of 5.

i) $\dfrac{5^6 \times 5^2}{5^4}$

$= \dfrac{5^8}{5^4} = 5^4$

ii) $\dfrac{5 \times 5^7}{(5^3)^2}$

$= \dfrac{5^8}{5^6} = 5^2$

> When no power is shown, the power is 1, i.e. $5 = 5^1$.

Negative Powers

- A negative power occurs when the power on the denominator (bottom number) of a fraction is higher than the power on the numerator (top number).

LEARN

$$\frac{1}{x^n} = x^{-n}$$

Write down $\dfrac{6^3}{6^5}$ in its simplest form.

$$\frac{6^3}{6^5} = \frac{\cancel{6} \times \cancel{6} \times \cancel{6}}{\cancel{6} \times \cancel{6} \times \cancel{6} \times 6 \times 6}$$

$$= \frac{1}{6 \times 6} = 6^{-2}$$

Write each of the following as a single power of 7.

a) $7^3 \div 7^9$

$= 7^{3-9}$

$= 7^{-6}$

b) $(7^{-2})^{-3}$

$= 7^{(-2 \times -3)}$

$= 7^6$

> Remember, $- \times - = +$

Simplify $(3x^2 y)^{-2}$

$= 3^{-2} \times (x^2)^{-2} \times y^{-2}$

$= \dfrac{1}{9} x^{-4} y^{-2}$

$= \dfrac{1}{9x^4 y^2}$

Quick Test

1. Simplify $2x^3 \times 3x^2$
2. Write $2^3 \times 2^5$ as a single power of 2.
3. What is the value of 3^{-3}?
4. Simplify $(x^{-2} y^5)^{-2}$

> **Key Words**
>
> index / indices
> root

Area and Volume 1

You must be able to:

- Calculate the perimeter and area of rectangles and triangles
- Calculate the perimeter and area of composite shapes
- Calculate the volume and surface area of a cuboid.

Rectangles

- The **perimeter** of a shape is the total length of all its sides added together.
- The **area** is the space enclosed within the perimeter.

> Area of a Rectangle (A) = Length (l) × Width (w)
> $$A = lw$$
>
> Perimeter of a Rectangle (P) = 2 × Length (l) + 2 × Width (w)
> $$P = 2l + 2w$$

Calculate the perimeter and area of the rectangle.

Area

$A = 5 \times 8$

$\quad = 40\text{cm}^2$

Perimeter

$P = 2 \times 5 + 2 \times 8$

$\quad = 10 + 16 = 26\text{cm}$

5cm

8cm

Triangles

> Area of a Triangle (A) = $\frac{1}{2}$ × Base (b) × Height (h)
>
> $$A = \frac{1}{2}bh$$

Work out the perimeter and area of the triangle.

Area

$A = \frac{1}{2}bh = \frac{1}{2} \times 4 \times 3$

$\quad = 6\text{cm}^2$

Perimeter

$P = 3 + 4 + 5$

$\quad = 12\text{cm}$

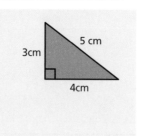

3cm

5 cm

4cm

> **Key Point**
>
> When calculating the area of a triangle, always use the perpendicular height.

Composite Shapes

- **Composite shapes** are made up of other shapes.
- To find the perimeter and area of composite shapes, break them down into their component shapes.

Calculate the perimeter and area of the shape below.

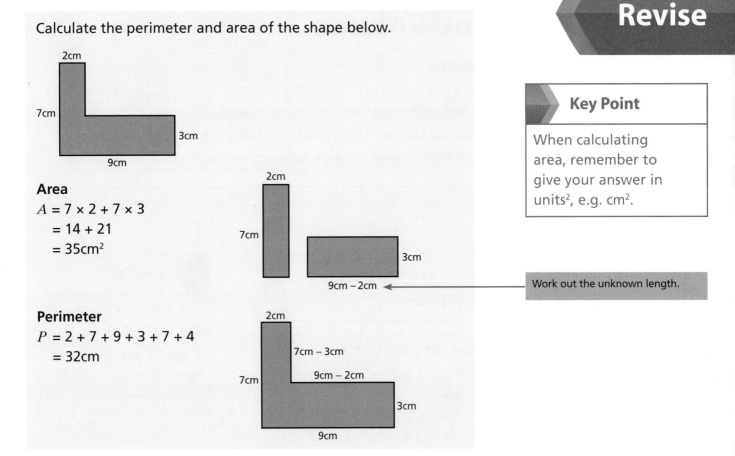

Work out the unknown length.

Area

$A = 7 \times 2 + 7 \times 3$

$\quad = 14 + 21$

$\quad = 35 \text{cm}^2$

Perimeter

$P = 2 + 7 + 9 + 3 + 7 + 4$

$\quad = 32 \text{cm}$

Key Point

When calculating area, remember to give your answer in units^2, e.g. cm^2.

Cuboids

LEARN

Volume of a Cuboid (V) = Length (l) × Width (w) × Height (h)

$$V = lwh$$

Surface Area of a Cuboid (SA) = $2lw + 2lh + 2hw$

Key Point

When calculating volume, remember to give your answer in units^3, e.g. cm^3.

Work out the volume and surface area of the cuboid.

$V = 8 \times 1.5 \times 2$

$\quad = 24 \text{cm}^3$

$SA = 2 \times 8 \times 1.5 + 2 \times 8 \times 2 + 2 \times 2 \times 1.5$

$\quad = 24 + 32 + 6$

$\quad = 62 \text{cm}^2$

Quick Test

1. A rectangle has a length of 4m and a width of 3m. Calculate the area and perimeter.
2. A triangle has a base of length 5cm and a perpendicular height of 4cm. Calculate the area.
3. A cuboid has a length of 3cm, a width of 3cm and a height of 2cm. Work out the volume and surface area of the cuboid.

Key Words

perimeter
area
composite shapes

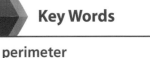

Area and Volume 2

You must be able to:

- Recall and use the formulae for the circumference and area of a circle
- Recall and use the formula for the area of a trapezium
- Recall and use the formulae for the volume and surface area of a prism
- Recall and use the formulae for the volume and surface area of a cylinder.

Circles

Circumference of a Circle (C) = $2\pi r$ or $C = \pi d$

Area of a Circle (A) = πr^2

> **Key Point**
>
> The symbol π represents the number **pi**.
>
> π can be approximated by 3.14 or $\frac{22}{7}$.

Work out the circumference and area of a circle with radius 9cm.
Give your answers to 1 decimal place.

Circumference

$C = 2 \times \pi \times 9$

$= 18 \times \pi$

$= 56.5$cm (to 1 d.p.)

Area

$A = \pi \times 9^2$

$= \pi \times 81$

$= 254.5$cm^2 (to 1 d.p.)

Trapeziums

The area of a **trapezium** is:

$$A = \tfrac{1}{2}(a + b)h$$

where a and b are the **parallel** sides and
h is the **perpendicular** height

> **Key Point**
>
> Perpendicular means 'at right angles'.
>
> Parallel means 'in the same direction and always the same distance apart'.

- This formula can be proved:

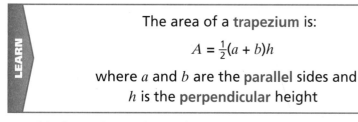

  - Two identical trapeziums fit together to make a parallelogram with base $a + b$ and height h
  - The area of the parallelogram is $(a + b)h$
  - Therefore, the area of each trapezium is $\tfrac{1}{2}(a + b)h$.

> **Key Point**
>
> The area of a parallelogram is: $A = bh$
>
>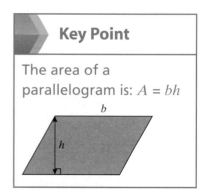

Work out the area of the trapezium.

$A = \tfrac{1}{2} \times (5 + 10) \times 4$

$= 30$cm^2

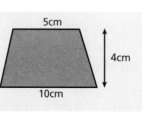

Prisms

- A right prism is a 3D shape that has the same **cross-section** running all the way through it.

> **LEARN**
> Volume of a Prism = Area of Cross-Section × Length

- The surface area is the sum of the areas of all the **faces**.

Work out the volume and surface area of the triangular prism.

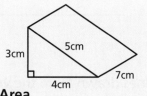

Volume
Area of the cross-section
$= \frac{1}{2} \times 3 \times 4 = 6 \text{cm}^2$
Volume $= 6 \times 7$
$= 42 \text{cm}^3$

Surface Area
Five faces:
Two triangular faces $= 6 + 6 = 12$
Base $= 4 \times 7 = 28$
Side $= 3 \times 7 = 21$
Slanted side $= 5 \times 7 = 35$
Total surface area $=$
$12 + 28 + 21 + 35 = 96 \text{cm}^2$

Cylinders

> **LEARN**
> Volume of a Cylinder $= \pi r^2 h$
>
> Surface Area of a Cylinder $= 2\pi r h + 2\pi r^2$

Work out the volume and the surface area of the cylinder. Give your answers in terms of π.

Volume
$V = \pi \times 4^2 \times 7$
$= 112\pi \text{cm}^3$

Surface Area
$SA = 2 \times \pi \times 4 \times 7 + 2 \times \pi \times 4^2$
$= 56\pi + 32\pi$
$= 88\pi \text{cm}^2$

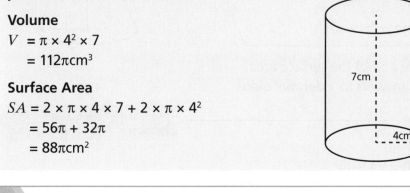

> **Key Point**
>
> A cylinder is just like any other right prism. To find the volume, you multiply the area of the cross-section (circular face) by the length of the cylinder.

> **Quick Test**
>
> 1. Calculate the volume and surface area of a cylinder with radius 4cm and height 6cm.
> 2. Work out the area of the trapezium.
>
>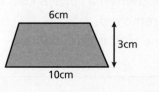
>
> 3. Calculate the circumference and area of a circle, diameter 7cm.

> **Key Words**
>
> trapezium
> parallel
> perpendicular
> cross-section
> face

Area and Volume 3

You must be able to:

- Find the volume of a pyramid
- Find the volume and surface area of a cone
- Find the volume of a frustum
- Find the volume and surface area of a sphere
- Find the area and volume of composite shapes.

Pyramids

- A **pyramid** is a 3D shape in which lines drawn from the **vertices** of the base meet at a point.

> **LEARN**
> Volume of a Pyramid = $\frac{1}{3}$ × Area of the Base × Height

> **Key Point**
>
> A pyramid is usually defined by the base, e.g. a square-based pyramid or a triangular-based pyramid.

Work out the volume of the square-based pyramid.

$V = \frac{1}{3} \times 9 \times 9 \times 7$

$= 189\text{cm}^3$

7cm

9cm   9cm

> You must use the perpendicular height to calculate volume.

Cones

- A **cone** is a 3D shape with a circular base that tapers to a single vertex.

> **LEARN**
> Volume of a Cone = $\frac{1}{3}\pi r^2 h$
>
> Surface Area of a Cone = $\pi r l + \pi r^2$

Work out **a)** the volume and **b)** the surface area of the cone. Give your answers to 1 decimal place.

6cm

4cm

a) $h = \sqrt{6^2 - 4^2}$

$h = \sqrt{20}$

$V = \frac{1}{3} \times \pi \times 4^2 \times \sqrt{20} = 74.9\text{cm}^3$

> First find the height using Pythagoras' Theorem (see p.84–85).

b) $SA = (\pi \times 4 \times 6) + (\pi \times 4^2) = 125.7\text{cm}^2$

- A **frustum** is the 3D shape that remains when a cone is cut parallel to its base and the top cone removed.
- The original cone and the smaller cone that is removed are always **similar**.

> **LEARN**
> Volume of a Frustum
> = Volume of Whole Cone – Volume of Top Cone

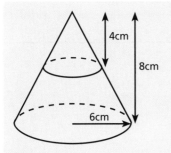

Calculate the volume of the frustum. Leave your answer in terms of π.

Radius of small cone = 3cm

$V = \frac{1}{3}(\pi \times 6^2 \times 8) - \frac{1}{3}(\pi \times 3^2 \times 4)$
$= 84\pi \text{cm}^3$

The two cones are similar with scale factor 2.

Spheres

- A **sphere** is a 3D shape that is round, like a ball. At every point, its surface is equidistant from its centre.

Volume of a Sphere = $\frac{4}{3}\pi r^3$

Surface Area of a Sphere = $4\pi r^2$

- A **hemisphere** is half of a sphere; a dome with a circular base.

Work out **a)** the volume and **b)** the surface area of the sphere. Leave your answers in terms of π.

a) $V = \frac{4}{3} \times \pi \times 6^3 = 288\pi \text{cm}^3$

b) $SA = 4 \times \pi \times 6^2 = 144\pi \text{cm}^2$

Composite Shapes

Calculate the area of the shaded region.

$A = (6 \times 7) - (\pi \times 1.5^2)$
$= 34.9 \text{cm}^2$ (to 1 d.p.)

Find the area of the rectangle and subtract the area of the circle.

Key Point

To find the volume of a composite shape, you must break the shape down.

Work out the volume of the shape.
Give your answers to 2 decimal places.

Volume of the Cylinder = $2.5^2 \times \pi \times 7.8 = 153.15 \text{cm}^3$

Volume of the Cone = $\frac{1}{3} \times \pi \times 2.5^2 \times 6.2 = 40.58 \text{cm}^3$

Total Volume = $153.15 + 40.58 = 193.73 \text{cm}^3$

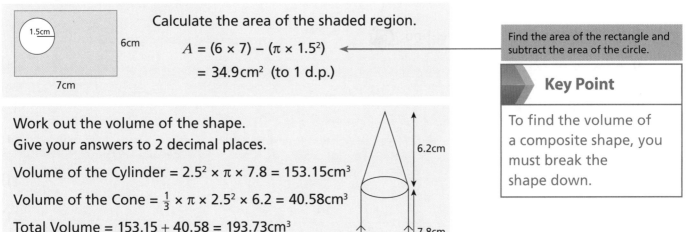

Quick Test

1. Work out the volume of a sphere with diameter 10cm.
2. Calculate the surface area of a cone with radius 3cm and perpendicular height 6cm.
3. Work out the volume of a square-based pyramid with side length 5cm and perpendicular height 8cm.
4. Calculate the surface area of a hemisphere with radius 6cm.

Key Words

pyramid
vertex / vertices
cone
frustum
similar
sphere
hemisphere

Review Questions

Number 1, 2 & 3

1 An adult theatre ticket costs £38.60 and a child ticket costs £12.76

  Work out the total cost for eight adult tickets and four child tickets. 📱 [2]

2 If 1.263 × 2.47 = 3.11961, work out: 📱

  a) 126.3 × 2.47 [1]

  b) 1.263 × 0.247 [1]

3 What is $(-3)^3$?
  Circle your answer. 📱

  −27   −9   9   27 [1]

4 In a restaurant, there are three different starters and eight different main meals.

  How many possible combinations are there when choosing a starter and a main meal? 📱 [1]

5 9, 13, 15, 27, 35, 100

  From the list of numbers above, find: 📱

  a) a factor of 72 [1]

  b) a multiple of 7 [1]

  c) a square number [1]

  d) a cube number [1]

  e) a prime number. [1]

6 Find a number (other than 1) that is a square number and also a cube number. 📱 [1]

7 Peter says $(2^3)^2 = (2^2)^3$

  Is Peter correct? [2]
  Write down a calculation to support your answer. 📱

8 Express 105 as the product of prime factors. 📱 [2]

> Total Marks _____ / 16

Basic Algebra & Factorisation and Formulae

1 There are k children in a room.
The number of children who wear glasses is g.

Write an expression in terms of k and g for the number of children who **do not** wear glasses. [1]

2 Simplify $7x - 2y + 5x - 3y$ [2]

3 Work out the value of the following expression when $a = -3$.

$$\frac{4a^2 - a^3}{a^4}$$

Circle your answer.

$\frac{1}{9}$ $\qquad\qquad -\frac{1}{9}$ $\qquad\qquad \frac{7}{9}$ $\qquad\qquad -\frac{7}{9}$ [1]

4 Solve $2x + 6 = 5x - 10$ [3]

5 Factorise $5ab - 3b^2c$ [1]

6 Solve $4(x - 3) = 10$ [3]

7 Solve $\frac{x}{3} + 3 = 1$ and circle your answer.

$x = 0$ $\qquad\qquad x = 12$ $\qquad\qquad x = -6$ $\qquad\qquad x = 8$ [1]

8 The shape below is a rectangle.

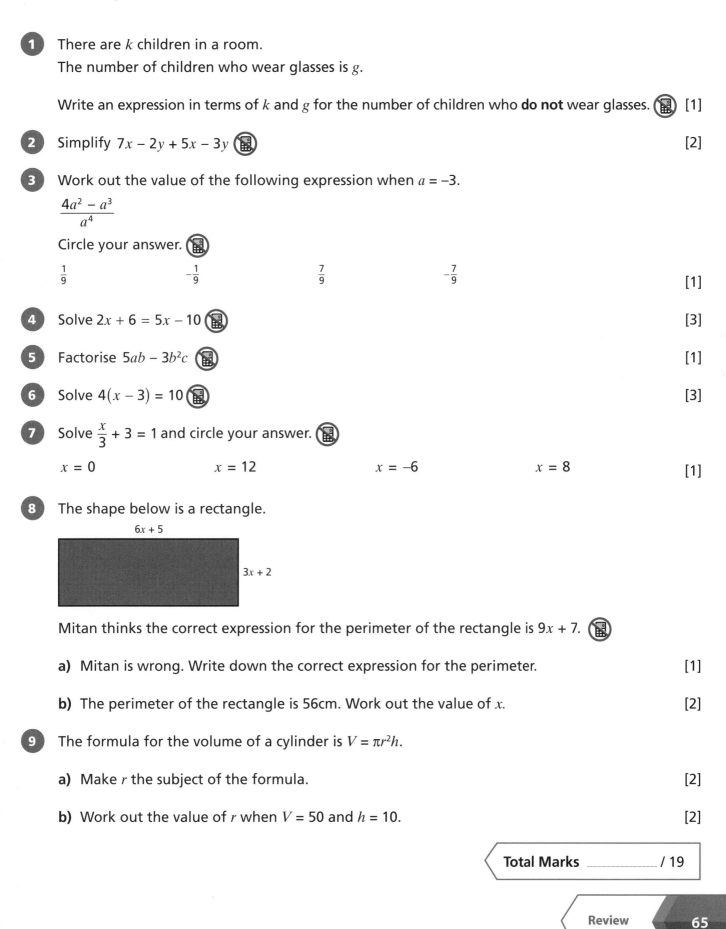

6x + 5

3x + 2

Mitan thinks the correct expression for the perimeter of the rectangle is $9x + 7$.

a) Mitan is wrong. Write down the correct expression for the perimeter. [1]

b) The perimeter of the rectangle is 56cm. Work out the value of x. [2]

9 The formula for the volume of a cylinder is $V = \pi r^2 h$.

a) Make r the subject of the formula. [2]

b) Work out the value of r when $V = 50$ and $h = 10$. [2]

Total Marks .... / 19

Review Questions

Ratio and Proportion

1 The square of the speed (v) at which a ball is thrown is directly proportional to the height (h) reached, so $v^2 = kh$.

A ball thrown at a speed of 10 metres per second reaches a height of 5 metres.

Calculate the constant of proportionality (k). [2]

2 £60 is divided in the ratio 5 : 7

What is the difference in value between the two shares? [3]

3 Simplify 6 hours : 4 minutes [2]

Total Marks .... / 7

Variation and Compound Measures

1 **a)** Calculate the distance travelled by a mouse moving at 1.5 metres per second for 1.5 seconds. [2]

b) The maximum speed that an antelope can run at is 55 miles per hour.
The maximum speed that a zebra can run at is 64 kilometres per hour.
Assume 8km = 5 miles

Which animal can run the fastest?
You **must** show your working. [2]

2 The density of aluminium is 2700kg/m³.

Work out the mass of a piece of aluminium that has a volume of 4.5m³. [2]

3 A woman with a weight (force) of 588 newtons is wearing stiletto heels. The point of each heel has an area of 0.0001m².

What is the pressure that each heel exerts on the ground? [2]

Total Marks .... / 8

Angles and Shapes 1 & 2

1 The three interior angles of a triangle are $y°$, $2y°$ and $3y°$.

Work out the size of the largest angle. [2]

2 A quadrilateral has angles of 80°, 123°, 165° and 40°.

Why is it impossible to draw this quadrilateral? [1]

3 Name all the quadrilaterals that can be drawn with four lines of length 5cm, 8cm, 5cm and 8cm. [2]

4 An aircraft flies from airport A on a bearing of 054° to airport B.

Work out the bearing that the aircraft must fly on in order to return to airport A. [1]

5 What is the size of one interior angle in an equilateral triangle? [1]

6 Calculate:

a) The sum of the interior angles in a regular octagon [2]

b) The size of one interior angle in a regular octagon. [1]

7 A map has a scale of 1cm : 3km. A lake on the map is 6.5cm long.

What is the actual length of the lake in kilometres? [1]

8 The angles in a triangle are $y + 5$, $3y - 16$ and $2y + 5$ degrees.

a) Write down an equation for the sum of the angles in the triangle.
Give your answer in its simplest form. [1]

b) Solve your equation to find the value of y. [1]

c) Work out the size of each angle in the triangle. [3]

9 Elaine calculates that the interior angle of a regular polygon is 158°.
Pauline says that Elaine has made a mistake.

Who is correct?
Give a reason for your answer and show all your working. [3]

Total Marks _____ / 19

Review Questions

Fractions

1. Work out $6 \div \frac{3}{4}$ 🚫🖩   [1]

2. What is half of a half of a quarter? 🚫🖩   [1]

3. A box contained 344 dog biscuits.

   Fluff ate one-eighth of them.

   How many biscuits were left? 🚫🖩   [2]

4. Lee spends £95 a week.

   If this is five-sixths of Lee's weekly wage, what does he earn in total each week? 🚫🖩   [2]

5. Change $\frac{5}{8}$ to a decimal. 🚫🖩   [1]

> **Total Marks** _____ / 7

Percentages 1 & 2

1. A coat costs £56. It is reduced by 35% in a sale.

   What is the sale price of the coat?   [2]

2. Richard says that 30% of £40 is the same as 40% of £30.

   Is he correct?   [2]
   Write down a calculation to support your answer.

3. A house is purchased for £215 000 and sold for £300 000.

   What is the percentage profit?
   Circle your answer.

   85%   39.5%   28.33%   75%   [1]

4. Write down 15cm as a percentage of 5m.   [1]

> **Total Marks** _____ / 6

Probability 1 & 2

1 The sides of a spinner are coloured red, blue, green and yellow.
The probability that the spinner will land on each colour is shown in the table below:

Colour	Red	Blue	Green	Yellow
Probability	0.3	0.3		0.1

a) Complete the table to work out the probability of landing on green. [1]

b) Estimate the number of times the spinner will land on the colour red if it is spun 50 times. [2]

c) Georgia and Annie are playing a game using the spinner.
Georgia suggests that she wins if the spinner lands on red or blue and Annie wins if the spinner lands on green or yellow.

Georgia thinks this is fair. Is she correct? Give a reason for your answer. [2]

2 A restaurant serves three courses: starters, mains and desserts. All customers have a main course.
The Venn diagram shows information about whether customers ordered starters and desserts.

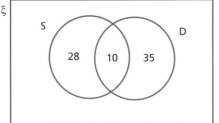

S = Starter
D = Dessert

The total number of customers served at lunchtime was 82.

Work out the number of customers who ordered neither a starter nor a dessert. [3]

3 Two events, A and B, are mutually exclusive: $P(A) = 0.3$ and $P(B) = 0.6$

a) Draw a Venn diagram to represent this information. [2]

b) Write down P(A and B) [1]

c) Work out P(A or B) [2]

4 Helen has ten yogurts. Four are vanilla flavoured and six are banana flavoured.
She takes a yogurt at random for breakfast on Monday.

Work out the probability that she takes a banana yogurt. [2]

Total Marks _____ / 15

Practice Questions

Number Patterns and Sequences 1 & 2

1 For each of the following sequences write down **i)** the next two terms in the sequence and **ii)** the term-to-term rule:

    a) 5, 8, 11, 14, 17, __, __   **[2]**

    b) 16, 12, 8, 4, 0, __, __   **[2]**

    c) 189, 63, 21, 7, $\frac{7}{3}$, __, __   **[2]**

2 Here is a sequence:

    20, 16, 12, 8, 4 ...

    Circle the expression for the nth term of the sequence.

    $20n - 4$ $4n + 20$ $4n - 24$ $24 - 4n$ **[1]**

3 Work out the first five terms in the sequence with the nth term $3n - 5$.   **[2]**

4 The first five terms of an arithmetic sequence are 14, 17, 20, 23, 26 ...

    a) Write an expression for the nth term of this sequence.   **[2]**

    b) Calculate the 100th term in this sequence.   **[1]**

5 **a)** A sequence of numbers is given as 5, 8, 12, 17 ...

       Write down the next two terms in the sequence.   **[2]**

    b) A second sequence of numbers is given as 2, 3, 2, 3, 2, 3 ...

       Write down the 100th term.   **[1]**

6 Here is a sequence of patterns made using matchsticks.

    a) Draw the next two patterns in the sequence.   **[2]**

    b) Write down the rule for the number of matchsticks required for pattern number n.   **[2]**

    c) Use the rule to work out how many matchsticks are required for pattern 100.   **[1]**

Total Marks .... / 20

Transformations, Constructions & Nets, Plans and Elevations

1 a) Plot the following points: A(2, 0) B(5, 0) C(5, 2) D(3, 2) E(3, 5) F(2, 5)
Join the points together and label the shape M.   [1]

b) Rotate shape M by 180° about the origin to form shape N.   [1]

c) Reflect shape N in the x-axis to form shape O.   [1]

d) Describe fully the single transformation that maps shape O to shape M.   [2]

2 Rectangle R has a width of 3cm and a length of 5cm.
It is enlarged by scale factor 3 to give rectangle T.

a) What is the area of rectangle T?   [2]

b) How many times bigger is the area of rectangle T than the area of rectangle R?   [2]

3 Describe how to construct an angle of 45°.   [2]

4 Describe the locus of points in the following:

a) A person sitting on the London Eye as it rotates around.   [1]

b) The seat of a moving swing.   [1]

c) The end of the minute hand on a clock moving for one hour.   [1]

d) The end of a moving see-saw.   [1]

5 The diagram represents a solid made from ten identical cubes.

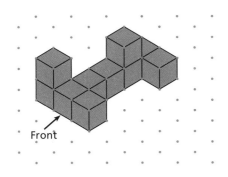

Front

On a squared grid, draw the:

a) Front elevation   **b)** Plan view.   [2]

6 Which of the following four nets produce cubes?
Circle your answers.

A   B   C   D   [1]

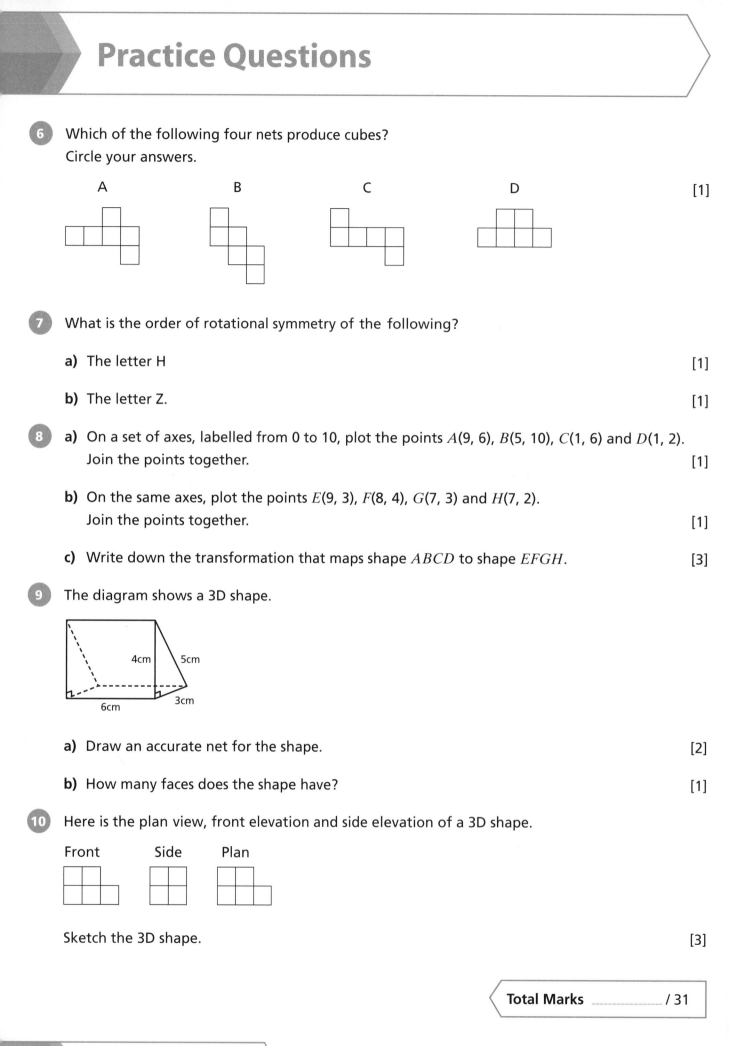

7 What is the order of rotational symmetry of the following?

a) The letter H [1]

b) The letter Z. [1]

8 a) On a set of axes, labelled from 0 to 10, plot the points $A(9, 6)$, $B(5, 10)$, $C(1, 6)$ and $D(1, 2)$.
Join the points together. [1]

b) On the same axes, plot the points $E(9, 3)$, $F(8, 4)$, $G(7, 3)$ and $H(7, 2)$.
Join the points together. [1]

c) Write down the transformation that maps shape $ABCD$ to shape $EFGH$. [3]

9 The diagram shows a 3D shape.

4cm   5cm
6cm   3cm

a) Draw an accurate net for the shape. [2]

b) How many faces does the shape have? [1]

10 Here is the plan view, front elevation and side elevation of a 3D shape.

Front   Side   Plan

Sketch the 3D shape. [3]

Total Marks .... / 31

Linear Graphs & Graphs of Quadratic Functions

1 A straight-line graph has the equation $y = 3x + 7$.

a) To plot the graph, you need to work out values of x and y for the equation.

Draw a flow diagram that can be used to work out values for x and y, where x is the input and y is the output. [2]

b) Use your flow diagram to complete the table below. [2]

x	−3	−2	−1	0	1	2	3
$y = 3x + 7$	−2				10		

c) Draw the graph of $y = 3x + 7$ for values of x from −3 to 3. [2]

2 **a)** Sketch the graph of $y = x^2 + 3x + 2$.
Clearly label the points where the graph crosses the axes. [3]

b) Write down the x-coordinate of the turning point of the graph. [1]

3 Draw the graph of $y = 4x − 2$ for values of x from −4 to 4. [2]

4 Draw the graph of the function with gradient 5 and y-intercept (0, 3) for values of x between −2 and 2. [2]

5 Write down the gradient and y-intercept of the graph with equation $y = 5 − 2x$. [1]

6 Draw the graph with equation $y = 3x^2 − 2x + 1$ for values of x from −3 to 3. [2]

7 Work out the equation of the line that joins the points (−3, 5) and (3, −1). [3]

8 Work out the equation of the line shown. [3]

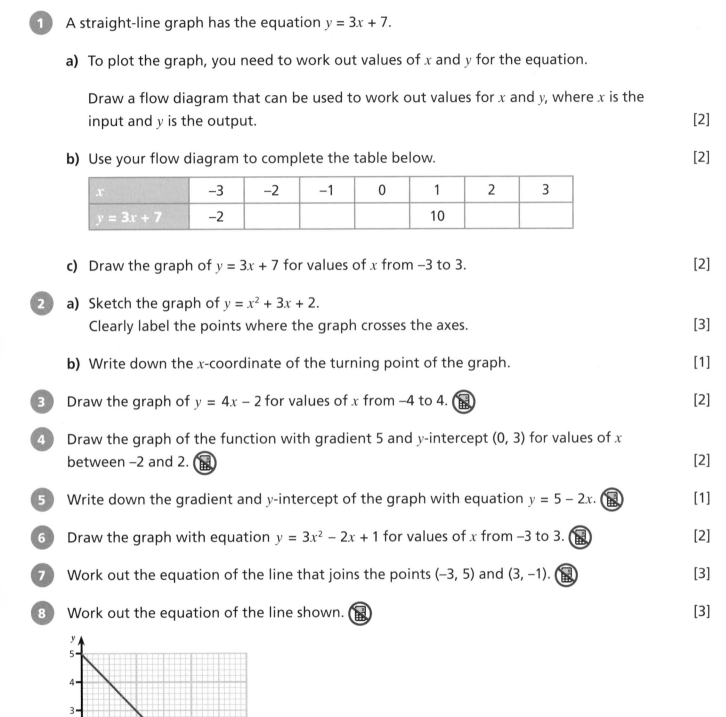

Total Marks .... / 23

Powers, Roots and Indices

1 Write down the following as a single power of 5: 📵

    a) $5^2 \times 5^3$ [1]

    b) $5^7 \div 5^3$ [1]

    c) $\dfrac{5^4 \times 5^2}{5^3}$ [2]

2 Simplify $(x^2)^4$
Circle your answer. 📵

    x^6 x^{16} x^8 x^{-2} [1]

3 Write down the following as a single power of 7:

    a) $7^2 \times 7^5 \times 7^{-3}$ [2]

    b) $\dfrac{7^4}{7^{-4}}$ [2]

    c) $\left(7^3\right)^{-2}$ [2]

4 Dan thinks that $2^2 \times 5^2 = 10^4$

Rebecca thinks $2^2 \times 5^2 = 100$

Who is correct?
Write down a calculation to support your answer.   [2]

5 Simplify $\left(9x^2\right)^{-2}$ [2]

6 Expand and simplify $x^{-3}\left(x^2 + x^3\right)$ [2]

7 Simplify $\left(3r^2p\right)^3$ 📵   [2]

8 Write down the value of:

    a) $2^3 \times 2^2$ [2]

    b) $3^3 \div 2$ [2]

    c) 4^{-3} [1]

    d) 5^0 [1]

Total Marks .... / 25

Area and Volume 1, 2 & 3

1 Work out:

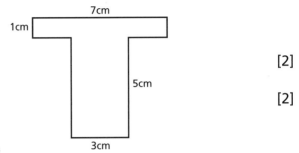

   a) The perimeter of the shape [2]

   b) The area of the shape. [2]

2 Here is a cuboid. The volume of the cuboid is 300cm³.

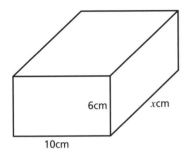

   Work out the value of x. [2]

3 Two identical circles sit inside a square of side length 6cm.

   Work out the area of the shaded region. [4]

4 A vase is made from two cylinders. The larger cylinder has a radius of 15cm.
   The total volume of the vase is 6000πcm³.
   The ratio of volumes of the smaller cylinder to the larger cylinder is 1 : 3.

   a) Calculate the height of the larger cylinder. [3]

   b) The height and radius of the smaller cylinder are equal.

   Work out the radius of the smaller cylinder. [3]

5 A cat's toy is made out of plastic. The top of the toy is a solid cone with radius 3cm and
   height 7cm. The bottom of the toy is a solid hemisphere. The base of the hemisphere and
   the base of the cone are the same size.

   Calculate the volume of plastic needed to make the toy. Give your answer in terms of π. [3]

   Total Marks _____ / 19

Uses of Graphs

You must be able to:

- Use the form $y = mx + c$ to identify parallel lines
- Interpret the gradient of a straight-line graph as a rate of change
- Recognise and interpret graphs that illustrate direct and inverse proportion.

Parallel Lines

- **Parallel** lines travel in the same direction and have the same gradient.

Write down the gradient of the line that is parallel to the line with the equation $y = 6x + 2$.

The line has a gradient of +6, so the line that is parallel to it also has a gradient of +6.

Write down the gradient of the line that is parallel to the line with the equation $y = 7 - 2x$.

The line has a gradient of –2, so the line that is parallel to it also has a gradient of –2.

Work out the equation of the line that goes through the point (2, 9) and is parallel to the line with equation $y = 7x + 10$.

$y = mx + c$
$y = 7x + c$ ← Substitute in $m = 7$.
$9 = (7 \times 2) + c$ ← Goes through the point (2, 9), so $x = 2$ when $y = 9$.
$c = -5$
The equation of the parallel line is $y = 7x - 5$.

> **Key Point**
>
> The gradient of a straight line in the form $y = mx + c$ is m.

Gradient of a Line

- The **rate of change** is the rate at which one quantity changes in relation to another.
- The gradient of a straight-line graph represents a rate of change – it describes how the variable on the y-axis changes when the variable on the x-axis is increased by 1.

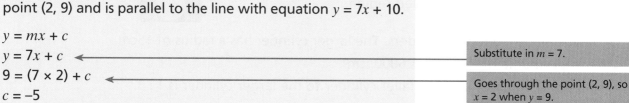

LEARN

$$\text{Gradient} = \frac{\text{Change in } y}{\text{Change in } x}$$

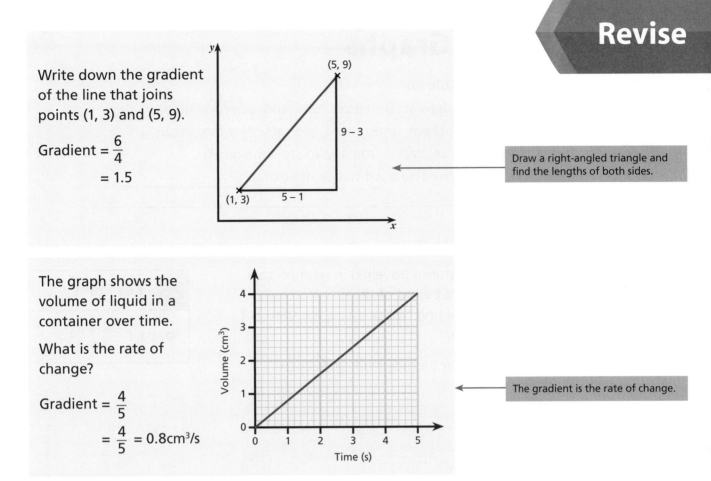

Write down the gradient of the line that joins points (1, 3) and (5, 9).

Gradient = $\frac{6}{4}$

= 1.5

Draw a right-angled triangle and find the lengths of both sides.

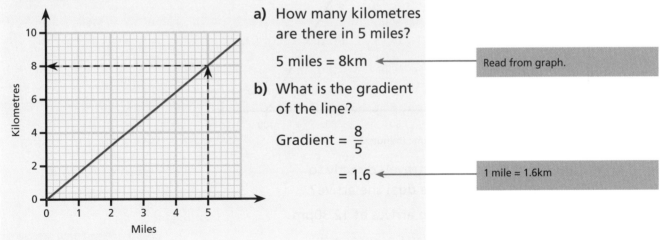

The graph shows the volume of liquid in a container over time.

What is the rate of change?

Gradient = $\frac{4}{5}$

= $\frac{4}{5}$ = 0.8cm³/s

The gradient is the rate of change.

Real-Life Uses of Graphs

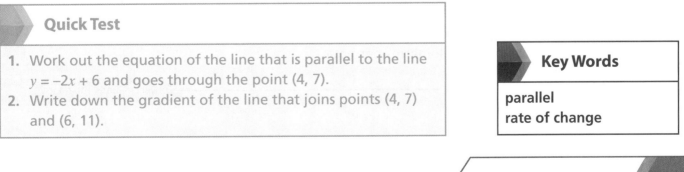

The graph below is the conversion graph between miles and kilometres.

a) How many kilometres are there in 5 miles?

5 miles = 8km

Read from graph.

b) What is the gradient of the line?

Gradient = $\frac{8}{5}$

= 1.6

1 mile = 1.6km

Quick Test

1. Work out the equation of the line that is parallel to the line $y = -2x + 6$ and goes through the point (4, 7).
2. Write down the gradient of the line that joins points (4, 7) and (6, 11).

Key Words

parallel
rate of change

Other Graphs

You must be able to:

- Recognise, draw and interpret cubic and reciprocal graphs
- Interpret distance–time graphs and velocity–time graphs
- Work out acceleration from a velocity–time graph
- Work out speed from a distance–time graph.

Distance–Time Graphs

- A **distance–time graph** shows distance travelled in relation to a fixed point (starting point) over a period of time.
- The gradient of a straight line joining two points is the speed of travel between those two points.

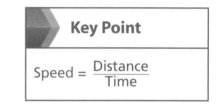

Key Point

$$\text{Speed} = \frac{\text{Distance}}{\text{Time}}$$

The graph below shows Val's car journey from St Bees to Cockermouth and back.

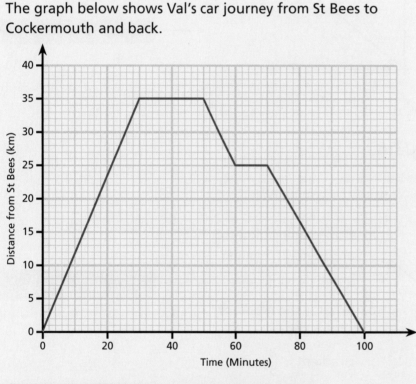

a) Val sets off at 12 noon and travels directly to Cockermouth. At what time does she arrive?

Val travels for 30 minutes so arrives at 12.30pm.

b) For how long does Val stop in Cockermouth?

20 minutes ◄———— This is represented by the horizontal line on the graph – where the distance does not change.

c) Val begins her journey home but stops to fill up with petrol.

Calculate the average speed of Val's journey from the petrol station to home in kilometres per hour.

$$\text{Speed} = \frac{\text{Distance}}{\text{Time}} = \frac{25}{0.5}$$ ◄———— Convert minutes into hours.

$$= 50\text{km/h}$$

Velocity–Time Graphs

- **Velocity** has both magnitude (size) and direction.
- The magnitude of velocity is called speed.
- The gradient of a straight line joining two points is the acceleration between those two points.
- Area Under Graph = Distance Travelled.

The graph below shows part of the journey of a car.

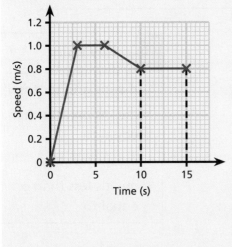

a) For how many seconds does the car decelerate?

4 seconds

b) What is the distance travelled in the last five seconds of the journey?

5 × 0.8 = 4m

c) What is the acceleration of the car in the first three seconds of the journey?

The acceleration is $\frac{1}{3}$ m/s²

Between 6 seconds and 10 seconds, there is a negative gradient so the car is decelerating.

This is the area of the rectangle under that part of the line.

This is the gradient of the line.

Other Graphs

- A **cubic function** is one that contains an x^3 term.
- A **reciprocal** function takes the form $y = \frac{a}{x}$.

Cubic Function

Reciprocal Function

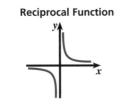

Quick Test

1. Plot the graph $y = 3x^3 - 5$ for values –2 to 2.
2. Below is a graph for the journey of a car.

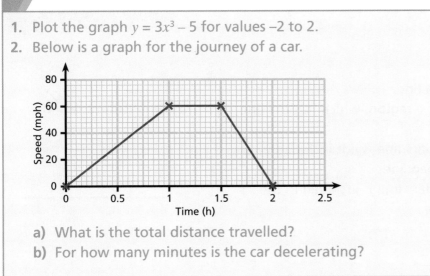

a) What is the total distance travelled?
b) For how many minutes is the car decelerating?

Key Words

distance–time graph
velocity
cubic function
reciprocal

Inequalities

You must be able to:

- Solve linear inequalities in one or two variables
- Represent solutions to inequalities on number lines or graphs.

Linear Inequalities

- The solution to an **inequality** can be shown on a number line.

  $\longleftarrow\!\circ$ means $x <$

  $\longleftarrow\!\bullet$ means $x \leqslant$

  $\circ\!\longrightarrow$ means $x >$

  $\bullet\!\longrightarrow$ means $x \geqslant$

Solve these inequalities and show the solutions on a number line:

a) $x + 3 > 4$

$x > 4 - 3$

$x > 1$

b) $2(x + 4) \leqslant 18$

$x + 4 \leqslant 9$

$x \leqslant 5$

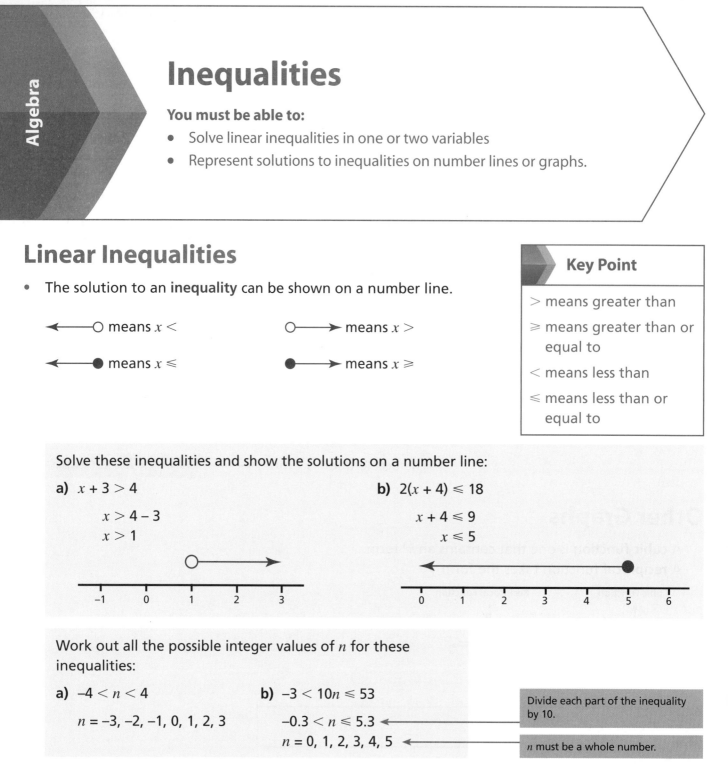

Work out all the possible integer values of n for these inequalities:

a) $-4 < n < 4$

$n = -3, -2, -1, 0, 1, 2, 3$

b) $-3 < 10n \leqslant 53$

$-0.3 < n \leqslant 5.3$ $\longleftarrow$

$n = 0, 1, 2, 3, 4, 5$ $\longleftarrow$

> Divide each part of the inequality by 10.

> n must be a whole number.

Graphical Inequalities

- The graph of the equation $y = 6$ is a line.
- The graph of the inequality $y > 6$ is a **region**, which has the line $y = 6$ as a boundary.
- For inequalities $>$ and $<$ the boundary line is **not included** in the solution and is shown as a **dashed line**.
- For inequalities $\geqslant$ and $\leqslant$ the boundary line is **included** in the solution and is shown as a **solid line**.

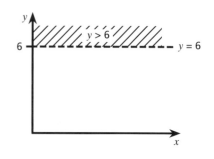

On a graph, show the region that satisfies $x \geqslant 0$, $x + y < 3$ and $y > 3x - 1$.

$x + y = 3$

$\quad y = 3 - x$

x	0	1	2
y	3	2	1

> Work out values of y for $x + y = 3$ and $y = 3x - 1$. You need three values.

$y = 3x - 1$

x	0	1	2
y	−1	2	5

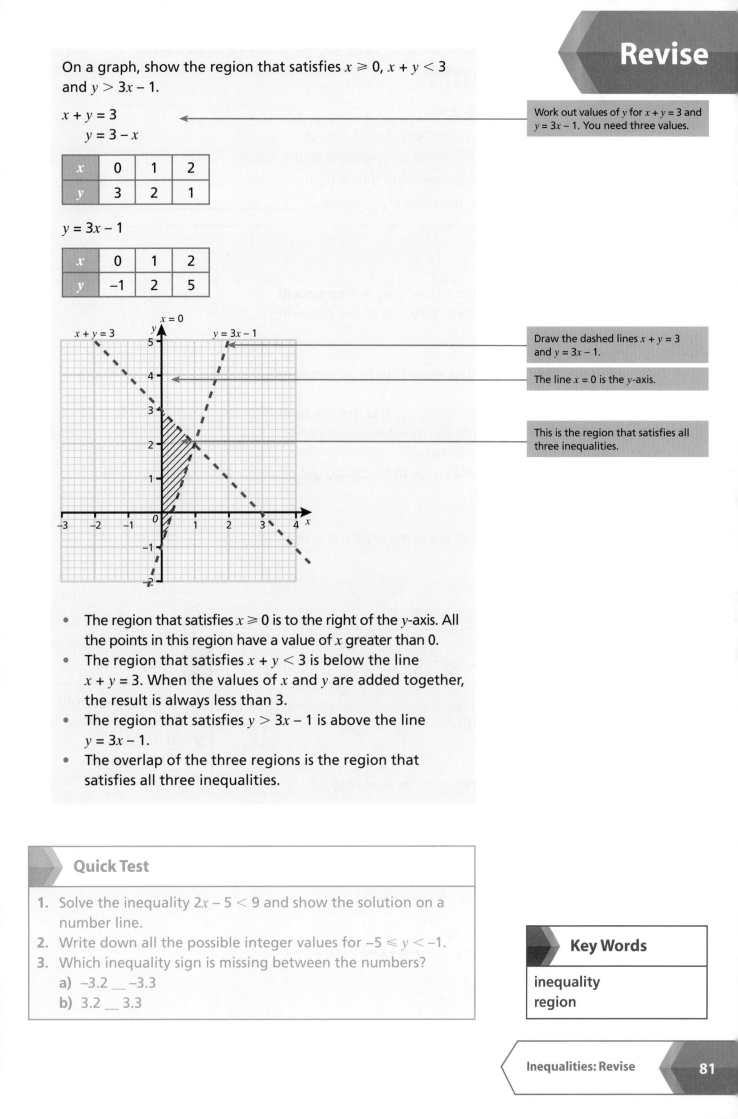

> Draw the dashed lines $x + y = 3$ and $y = 3x - 1$.

> The line $x = 0$ is the y-axis.

> This is the region that satisfies all three inequalities.

- The region that satisfies $x \geqslant 0$ is to the right of the y-axis. All the points in this region have a value of x greater than 0.
- The region that satisfies $x + y < 3$ is below the line $x + y = 3$. When the values of x and y are added together, the result is always less than 3.
- The region that satisfies $y > 3x - 1$ is above the line $y = 3x - 1$.
- The overlap of the three regions is the region that satisfies all three inequalities.

Quick Test

1. Solve the inequality $2x - 5 < 9$ and show the solution on a number line.
2. Write down all the possible integer values for $-5 \leqslant y < -1$.
3. Which inequality sign is missing between the numbers?
   a) −3.2 ___ −3.3
   b) 3.2 ___ 3.3

Key Words

inequality
region

Congruence and Geometrical Problems

You must be able to:

- Identify congruent and similar shapes
- State the criteria that congruent triangles satisfy
- Solve problems involving similar figures
- Understand geometrical problems.

Congruent Triangles

- If two shapes are the same size and shape, they are **congruent**.
- Two triangles are congruent if they satisfy one of the following four criteria:
  - SSS – three sides are the same
  - SAS – two sides and the **included angle** (the angle between the two sides) are the same
  - ASA – two angles and one corresponding side are the same
  - RHS – there is a right angle, and the hypotenuse and one other corresponding side are the same.
- Sometimes angles or lengths of sides have to be calculated before congruency can be proved.

State whether these two triangles are congruent and give a reason for your answer.

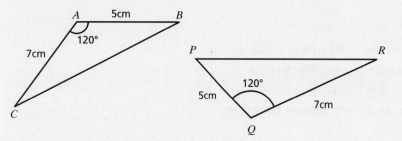

Angle CAB = Angle PQR (given)
$AC = QR$ (given)
$AB = PQ$ (given)
Triangles ABC and PQR are congruent because they satisfy the criteria SAS.

> **Key Point**
>
> Congruent shapes can be reflected, rotated or translated and remain congruent.

Similar Triangles

- **Similar** figures are identical in shape but can differ in size.
- In similar triangles:
  - corresponding angles are identical
  - lengths of corresponding sides are in the same ratio $y : z$
  - the area ratio = $y^2 : z^2$
  - the volume ratio = $y^3 : z^3$.

Triangles AED and ABC are similar.

Calculate a) AC and b) DC.

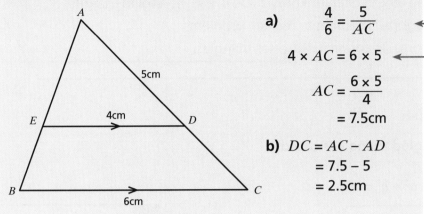

a) $\dfrac{4}{6} = \dfrac{5}{AC}$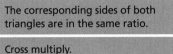

The corresponding sides of both triangles are in the same ratio.

$$4 \times AC = 6 \times 5$$

Cross multiply.

$$AC = \dfrac{6 \times 5}{4}$$
$$= 7.5\text{cm}$$

b) $DC = AC - AD$
$$= 7.5 - 5$$
$$= 2.5\text{cm}$$

Geometrical Problems

- Congruency and similarity are used in many geometric proofs.

Prove that the base angles of an isosceles triangle are equal.

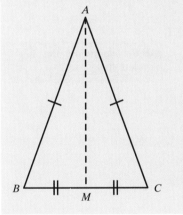

Given $\triangle ABC$ with $AB = AC$
Let M be the midpoint of BC
Join AM
$AB = AC$ (given)
$BM = MC$ (from construction)
$AM = AM$ (common side)
$\triangle ABM$ and $\triangle ACM$ are congruent (SSS)
So, angle ABC = angle ACB

Quick Test

1.

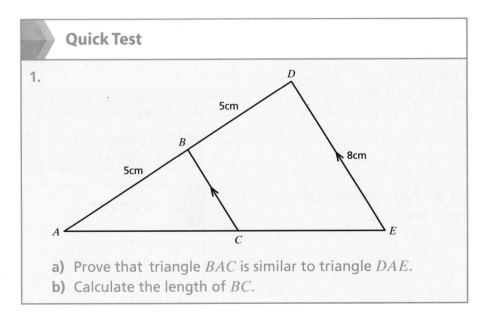

a) Prove that triangle BAC is similar to triangle DAE.
b) Calculate the length of BC.

Right-Angled Triangles 1

You must be able to:

- Recall and use the formula for Pythagoras' Theorem
- Calculate the length of an unknown side in a right-angled triangle
- Apply Pythagoras' Theorem to real-life problems
- Use Pythagoras' Theorem in isosceles triangles.

Pythagoras' Theorem

- The longest side (c) of a right-angled triangle is called the **hypotenuse**.
- **Pythagoras' Theorem** states that $a^2 + b^2 = c^2$.

LEARN

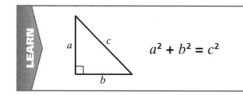 $a^2 + b^2 = c^2$

> **Key Point**
>
> a^2 means $a \times a$ **not** $2 \times a$

Calculating Unknown Sides

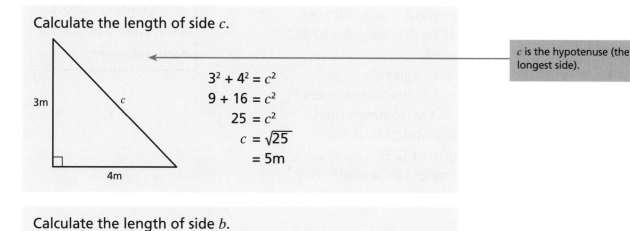

Calculate the length of side c.

$$3^2 + 4^2 = c^2$$
$$9 + 16 = c^2$$
$$25 = c^2$$
$$c = \sqrt{25}$$
$$= 5m$$

c is the hypotenuse (the longest side).

Calculate the length of side b.

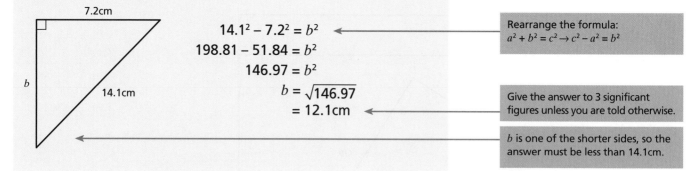

$$14.1^2 - 7.2^2 = b^2$$
$$198.81 - 51.84 = b^2$$
$$146.97 = b^2$$
$$b = \sqrt{146.97}$$
$$= 12.1cm$$

Rearrange the formula:
$a^2 + b^2 = c^2 \rightarrow c^2 - a^2 = b^2$

Give the answer to 3 significant figures unless you are told otherwise.

b is one of the shorter sides, so the answer must be less than 14.1cm.

- The following length combinations are **Pythagorean triples**. They regularly appear in right-angled triangles:
  - (3, 4, 5)
  - (6, 8, 10)
  - (5, 12, 13)
  - (7, 24, 25).

> **Key Point**
>
> Memorise the Pythagorean triples to help identify unknown sides quickly.

Real-Life Problems

A boat sails 15km due north and then 10km due east.

How far is the boat from its starting point?
Give your answer to 3 decimal places.

$$15^2 + 10^2 = c^2$$
$$225 + 100 = c^2$$
$$325 = c^2$$
$$c = \sqrt{325} = 18.028\text{km}$$

A sketch makes it clear that you are looking for the hypotenuse (c) of a right-angled triangle.

A four-metre ladder leans against a tree.
It reaches three metres up the side of the tree.

Calculate how far the base of the ladder is from the bottom of the tree.

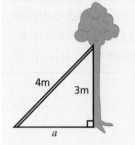

$$4^2 - 3^2 = a^2$$
$$16 - 9 = a^2$$
$$7 = a^2$$
$$a = \sqrt{7}$$
$$= 2.65\text{m}$$

You are looking for one of the shorter sides of a right-angled triangle.

Isosceles Triangles

Calculate the height of the isosceles triangle ABD.

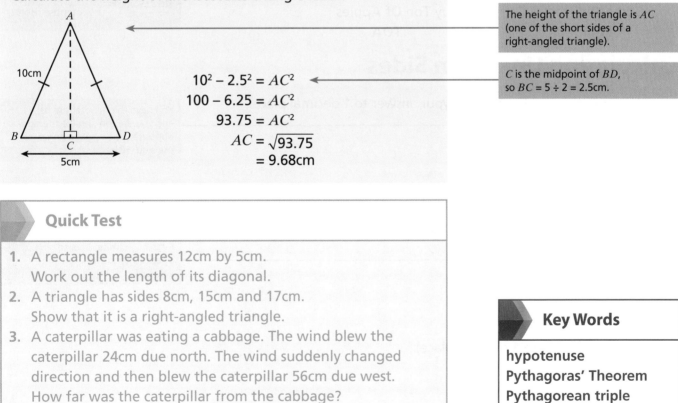

$$10^2 - 2.5^2 = AC^2$$
$$100 - 6.25 = AC^2$$
$$93.75 = AC^2$$
$$AC = \sqrt{93.75}$$
$$= 9.68\text{cm}$$

The height of the triangle is AC (one of the short sides of a right-angled triangle).

C is the midpoint of BD, so $BC = 5 \div 2 = 2.5$cm.

Quick Test

1. A rectangle measures 12cm by 5cm.
   Work out the length of its diagonal.
2. A triangle has sides 8cm, 15cm and 17cm.
   Show that it is a right-angled triangle.
3. A caterpillar was eating a cabbage. The wind blew the caterpillar 24cm due north. The wind suddenly changed direction and then blew the caterpillar 56cm due west. How far was the caterpillar from the cabbage?

Key Words

hypotenuse
Pythagoras' Theorem
Pythagorean triple

Right-Angled Triangles 2

You must be able to:

- Recall and use the trigonometric ratios
- Calculate unknown lengths and angles using trigonometry
- Recall the exact trigonometric values for certain angles without using a calculator.

The Trigonometric Ratios

- Unknown sides or angles in right-angled triangles can be calculated using the **trigonometric ratios**: **sine**, **cosine** and **tangent**.
- The symbol θ (**theta**) is used to represent an unknown angle.

LEARN

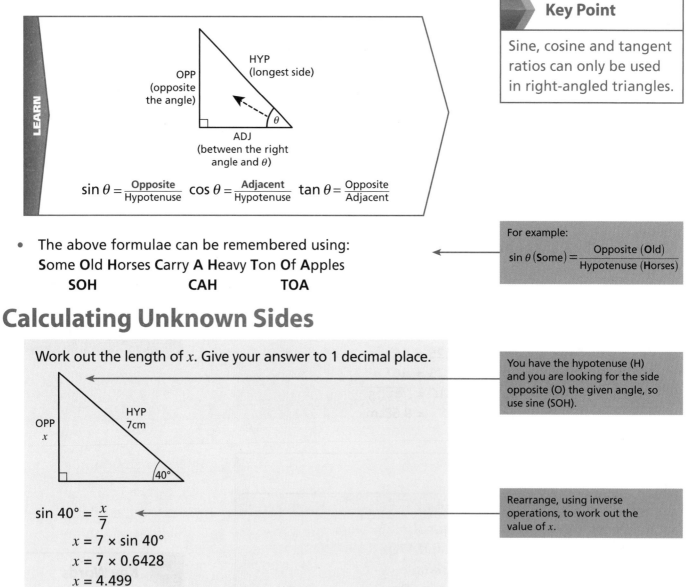

$$\sin \theta = \frac{\text{Opposite}}{\text{Hypotenuse}} \quad \cos \theta = \frac{\text{Adjacent}}{\text{Hypotenuse}} \quad \tan \theta = \frac{\text{Opposite}}{\text{Adjacent}}$$

> **Key Point**
>
> Sine, cosine and tangent ratios can only be used in right-angled triangles.

- The above formulae can be remembered using:
  Some **O**ld **H**orses **C**arry **A** **H**eavy **T**on **O**f **A**pples
  SOH   **CAH**   **TOA**

For example:
$$\sin \theta \,(\textbf{S}\text{ome}) = \frac{\text{Opposite }(\textbf{O}\text{ld})}{\text{Hypotenuse }(\textbf{H}\text{orses})}$$

Calculating Unknown Sides

Work out the length of x. Give your answer to 1 decimal place.

You have the hypotenuse (H) and you are looking for the side opposite (O) the given angle, so use sine (SOH).

$$\sin 40° = \frac{x}{7}$$
$$x = 7 \times \sin 40°$$
$$x = 7 \times 0.6428$$
$$x = 4.499$$
$$x = 4.5\text{cm (to 1 decimal place)}$$

Rearrange, using inverse operations, to work out the value of x.

The diagram shows a lookout tower.
Ann is standing 30m from the base of the tower.

If the angle of elevation to the top of the tower is 50°, calculate the height of the tower. Give your answer to 1 decimal place.

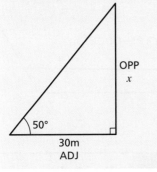

$$\tan 50° = \frac{x}{30}$$

$$x = 30 \times \tan 50°$$

$$x = 30 \times 1.1918$$

$$= 35.8\text{m (to 1 decimal place)}$$

You have the adjacent (A) side and you are looking for the side opposite (O) the given angle, so use tangent (TOA).

> **Key Point**
>
> Angles of elevation are angles above the horizontal, e.g. the angle from the ground to the top of a tower.

Calculating Unknown Angles

Work out the size of angle θ.
Give your answer to the nearest degree.

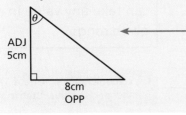

$$\tan \theta = \frac{8}{5} = 1.6$$

$$\tan^{-1} 1.6 = 57.99°$$

$$\theta = 58°$$

You have the adjacent (A) side and the opposite side (O), so use tangent (TOA).

On your calculator press
| SHIFT | tan | 1 | . | 6 |

EFG is a right-angled triangle. Angle $EFG = 90°$.
$FG = 5$cm and $EG = 8$cm.

Calculate angle EGF. Give your answer to 1 decimal place.

$$\cos \theta = \frac{5}{8} = 0.625$$

$$\cos^{-1} 0.625 = 51.3°$$

EG is the hypotenuse (H) as it is opposite the right angle and FG is the side adjacent (A) to the unknown angle, so use cosine (CAH).

Trigonometric Values to Learn

> **Key Point**
>
> Make sure your calculator is in degree mode when using the trigonometric ratios.

	sin	cos	tan
0°	0	1	0
30°	$\frac{1}{2}$	$\frac{\sqrt{3}}{2}$	$\frac{1}{\sqrt{3}}$
45°	$\frac{1}{\sqrt{2}}$	$\frac{1}{\sqrt{2}}$	1
60°	$\frac{\sqrt{3}}{2}$	$\frac{1}{2}$	$\sqrt{3}$
90°	1	0	infinity

> **Key Words**
>
> trigonometric ratios
> sine
> cosine
> tangent
> theta
> opposite
> adjacent

> **Quick Test**
>
> 1. An 8m ladder leans against a vertical wall. The base of the ladder is 3.5m from the wall. Calculate the angle between the top of the ladder and the wall. Give your answer to one decimal place.
> 2. A ship sails 14km on a bearing of 035°. How far north has the ship travelled? Give your answer to the nearest kilometre.

Statistics 1

You must be able to:

- Understand and identify different types of data
- Construct tally charts and frequency tables
- Construct pie charts, bar charts and line graphs to represent data.

Data and Data Collection

- Data can be **qualitative** (non-numerical) or **quantitative** (numerical).
- Numerical data can be **discrete** or **continuous**.
- A **tally chart** is a quick way of collecting data.
- When you use a tally chart to collect data, it is already grouped.
- To produce a **frequency table**, add a column to your tally chart, containing the total for each group.

> **Key Point**
>
> Discrete data takes certain values in a given range. Continuous data can take any value in a given range.

Louisa records the colour of the cars that pass by her house during a one-hour time period. She constructs a tally chart and frequency table to show the data.

The last car she sees is green. Add this to the tally chart and complete the frequency column.

Car colour is qualitative data. The number of cars is quantitative data.

Colour	Tally	Frequency
Red	JHT JHT JHT JHT JHT	25
Blue	JHT JHT JHT JHT I	21
Silver	JHT JHT IIII	14
White	JHT JHT JHT IIII	19
Other	JHT JHT I	11

Each line in the tally represents one car. Grouping them in sets of five makes them easy to count.

The frequency is the total number of cars for each group.

Statistical Diagrams

- The type of diagram you use will be dictated by the data and what aspect of the data you want to look at.
- **Bar charts** and vertical line graphs can be used to compare frequencies.

Draw a bar chart and vertical line graph to represent the data on car colours (above).

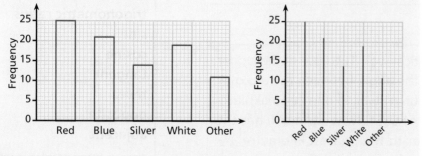

- **Pie charts** show proportion (but not exact frequencies).

Construct a pie chart to represent the data on car colours.

Colour	Frequency	Angle on Pie Chart
Red	25	25 × 4 = 100°
Blue	21	21 × 4 = 84°
Silver	14	14 × 4 = 56°
White	19	19 × 4 = 76°
Other	11	11 × 4 = 44°
Total	**90**	**360°**

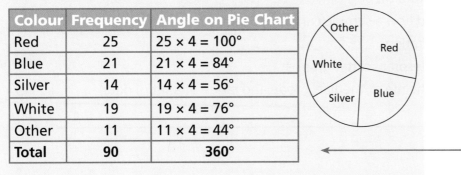

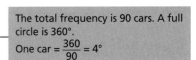

The total frequency is 90 cars. A full circle is 360°.
One car = $\frac{360}{90}$ = 4°

- A **line graph** is used to show changes in trends over time.

The graph shows ice cream sales for Shelby's Ices for a one-year period.

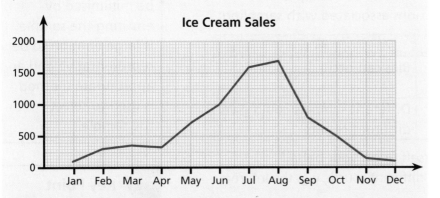

Which two months had the highest sales? Suggest a reason for this.

July and August had the highest sales.

Sales of ice cream are higher in these months because it is the summer, so temperatures are higher and more people are on holiday.

Quick Test

1. Jasmine surveyed the pupils in her class to find out their shoe size. Her results are shown below:

Shoe Size	Frequency
4	8
5	15
6	6
7	1

a) Draw a bar chart to represent her results.
b) Draw a pie chart to represent her results.
c) Which diagram do you think is most suitable for this data and why?

Key Words

qualitative
quantitative
discrete
continuous
tally chart
frequency table
bar chart
pie chart
line graph

Statistics 2

You must be able to:

- Understand how a sample can be used to represent a population and its limitations
- Use and interpret scatter graphs
- Calculate the mean, median, mode and range of a set of data.

Sampling

- A **population** is a collection of individuals or items.
- For a large population, collecting information from all members is not practical and so information is obtained from a proportion of the population, referred to as a **sample**.
- There are advantages and limitations associated with sampling:

Advantages	Limitations
Quicker and cheaper than investigating whole population	Bias can occur
Can be impractical to investigate whole population	Different samples could give different results

> **Key Point**
>
> Limitations can be minimised by ensuring the sample is large enough to be representative and an appropriate method is used to select the sample.

- **Primary data** is collected by yourself or on your behalf.
- **Secondary data** is collected from a different source, e.g. the Internet.

Statistical Measures

- To compare data sets you should compare:
  - a measure of average: **mean, median** or **mode**
  - a measure of spread, i.e. the **range** – the difference between the largest and smallest value in the data set.

> **Key Point**
>
> An outlier is a value within a data set that is significantly smaller or larger than the other values.

The data shows the ages of the competitors in a swimming race:

18, 26, 19, 21, 18, 18, 48, 20, 20

a) Identify the outlier.

   The outlier is 48.

b) Remove the outlier from the data and calculate:

   i) The mean age

   $$\frac{18 + 26 + 19 + 21 + 18 + 18 + 20 + 20}{8} = \frac{160}{8} = 20$$

   Mean = $\frac{\text{Sum of all Values}}{\text{Number of Values}}$

   ii) The median age

   18, 18, 18, ⑲, ⑳, 20, 21, 26

   The median is 19.5

   Write the data in order, from smallest to largest. The median is the middle value.

   iii) The mode age.

   18

   The mode is the most common value.

The table shows the time taken by students, in minutes, to complete a mathematical puzzle.

Time Taken to Complete Puzzle (min)	Frequency (f)	Midpoint (x)	fx
$0 < t \leqslant 2$	8	1	8
$2 < t \leqslant 4$	12	3	36
$4 < t \leqslant 6$	10	5	50
$6 < t \leqslant 8$	5	7	35
	$\Sigma f = 35$		$\Sigma fx = 129$

a) Write down the modal class.

The modal class is $2 < t \leqslant 4$

b) Work out which group contains the median.

$\dfrac{35 + 1}{2} = $ 18th value

The median (18th value) is in the group $2 < t \leqslant 4$

c) Estimate the mean time taken.

Mean $= \dfrac{129}{35} = 3.69$ minutes (to 2 d.p.)

Key Point

Σ means 'sum of'.

The modal class has the highest frequency.

The median is the middle value.

There are 35 students.

$$\text{Mean} = \frac{\Sigma fx}{\Sigma f}$$

Scatter Graphs

- **Scatter graphs** or scatter diagrams are used to investigate the relationship between two variables.
- If a linear relationship exists, a line of best fit can be drawn. This can be used to make predictions.
- A prediction taken from the line of best fit within the data range is reliable.
- A prediction taken from the line of best fit outside the data range is less reliable.
- **Positive correlation** – as one variable increases, the other variable increases.
- **Negative correlation** – as one variable increases, the other variable decreases.

Positive Correlation

As temperature increases, ice-cream sales increase.

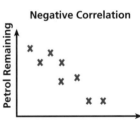

Negative Correlation

As distance travelled increases, petrol remaining decreases.

Quick Test

1. The table below shows the age of children who attend a reading group at the library. Calculate **a)** the mean **b)** the median **c)** the mode and **d)** the range of the data.

Age	6	7	8	9
Frequency	5	10	6	12

Key Words

population
sample
primary data
secondary data
mean
median

mode
range
scatter graph
positive correlation
negative correlation

Number Patterns and Sequences 1 & 2

1 The first term that the following two sequences have in common is 17.

8, 11, 14, 17, 20 ...

1, 5, 9, 13, 17 ...

Work out the next term that the two sequences have in common.
You must show your working. [2]

2 Regular pentagons of side length 1cm are joined together to make a pattern.

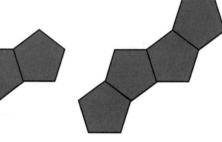

a) Use the patterns to complete the table below.

Pattern Number	Perimeter (cm)
1	
2	
3	
4	
60	
n	

[2]

b) What is the maximum number of pentagons that could be used to give a perimeter less than 1500cm? [2]

3 Write down the first three terms in the sequence with the nth term $n^2 - 6$. [2]

4 Write down the next two terms in the sequence below:

4, 6, 10, 18, 34 ... [2]

Total Marks _____ / 10

Transformations, Constructions & Nets, Plans and Elevations

1 Three points $X(5,1)$, $Y(3, 5)$, and $Z(1, 2)$ are reflected in the y-axis.

   a) Give the new coordinates of the three points. [3]

   b) The original points X, Y, and Z are rotated 90° about (0, 0) in a clockwise direction.

     Give the coordinates of the three points in their new positions. [3]

2 A rectangle (C) measures 3cm by 5cm. Each length of rectangle (C) is enlarged by scale factor 3 to form a new rectangle (D).

  What is the ratio of the area of rectangle C to rectangle D? [3]

3 A cuboid (C) measures 3cm by 4cm by 5cm. Each length of cuboid C is enlarged by scale factor 3 to form a new cuboid (D).

  What is the ratio of the volume of cuboid C to cuboid D? [3]

4 On a 6 × 6 grid, plot the points $A(3, 2)$, $B(1, 3)$, $C(0, 6)$ and $D(2, 5)$.

  Reflect each point in the line that joins (3, 0) to (3, 6) and write down the coordinates of points A', B', C' and D' in the image produced. [4]

5 Describe the locus of points for the following:

   a) The path of a rocket for the first three seconds after take-off. [1]

   b) A point just below the handle on an opening door. [1]

   c) The central point of a bicycle wheel as the bicycle travels along a level road. [1]

   d) The end of a pendulum on a grandfather clock. [1]

6 Describe the plan view of a cube measuring 4cm by 4cm by 4cm. [1]

7 **a)** Construct a triangle, DEF, where $DE = 8$cm, $EF = 7$cm and $DF = 3$cm. [2]

   b) By accurate measurement, find the size of angle FDE. [1]

   c) Construct the bisector of angle FED. [2]

8 The photograph shows a World War II Lancaster Bomber.

Sketch:

a) the side elevation of the Lancaster Bomber [2]

b) the front elevation of the Lancaster Bomber [2]

c) the plan view of the Lancaster Bomber. [2]

9 Below is a 3D shape made up of eight cubes.

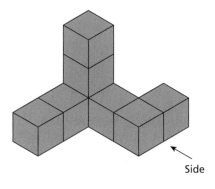

Side

a) Draw the plan view. [2]

b) Draw the side elevation. [2]

10 a) Here is the top half of a word. The dashed line is the line of symmetry.

Write down the word. [1]

b) Give an example of another three or four-letter word, through which a horizontal line of symmetry can be drawn. [1]

Total Marks _____ / 38

Linear Graphs & Graphs of Quadratic Functions

1 Work out the equation of the line that joins the points $\left(\frac{2}{3}, 8\right)$ and $\left(\frac{5}{6}, 5\right)$. [3]

2 Work out the equation of the line drawn below. [3]

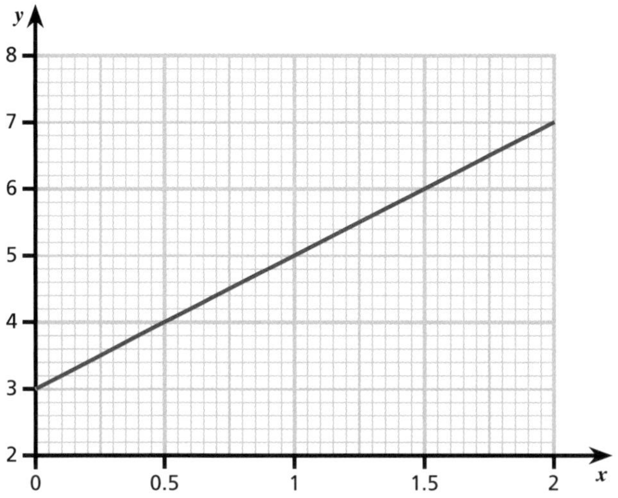

3 Sketch the graph of the function $y = x^2 + 4x + 3$, clearly stating the roots and the coordinates of the turning point. [3]

4 The equation of a line is $4y = 3x + 1$.

Work out the gradient and y-intercept of the line. [2]

5 A curve has the equation $y = x^2 + ax + b$.
The curve crosses the x-axis at the points $(-7, 0)$ and $(1, 0)$.

a) Work out the values of a and b. [3]

b) Work out the x-coordinate of the turning point. [1]

6 a) Sketch the graph $y = \frac{1}{x}$. [1]

b) On the same axes sketch the graph $y = -\frac{1}{x}$. [1]

7 Sketch the graph of $y = x^3$. [1]

Total Marks _____ / 18

Powers, Roots and Indices

1 Simplify $(2ab^{-5})^{-3} \times (3a^{-2}b^3)^2$ [3]

2 Amber states that $(x^{-2})^3 = \frac{1}{x^6}$

Is Amber's statement true or false?
Write down a calculation to support your answer. [2]

3 Write down the following as a single power of 3:

a) $3^2 \times 3^4$ [1]

b) $3^9 \div 3^6$ [1]

c) $(3^2)^3$ [1]

4 Write down the following as a single power of 5:

a) $5^4 \times 5^5 \times 5^{-2}$ [2]

b) $\frac{5^5}{5^{-4}}$ [2]

c) $(5^3)^{-6}$ [2]

5 Kirsten thinks that $3^2 + 3^3 = 3^5$
Darcey thinks $3^2 + 3^3 = 36$

Who is correct?
Write down a calculation to support your answer. [2]

6 Simplify $(16x^3)^{-2}$ [2]

7 Expand and simplify $x^5(x^{-3} + x^3)$ [2]

Total Marks _____ / 20

Area and Volume 1, 2 & 3

1 **a)** Work out the volume of the triangular prism. [2]

b) A cube has the same volume as the triangular prism.

Work out the total length of all the edges of the cube. [3]

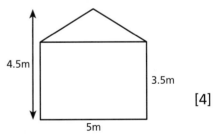

2 The numerical values of the area and circumference of a circle are equal.

Work out the radius of this circle. [2]

3 The volume of the trapezoid is 900cm³.
All measurements are in centimetres.

Work out the value of x. [4]

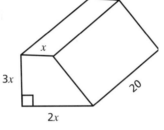

4 The surface area of a sphere is 75cm². Work out the length of the radius. [3]

5 Here is a triangle.
The area of the triangle is 7.5cm².

Work out the value of x. [3]

6 John is planning to paint the front of his house. He needs to estimate how much paint he should buy. He does this by calculating the area of the front of the house, including all windows and doors.

The diagram shows John's house.

If each tin of paint will cover 11m², work out an estimate of the number of tins that John needs to buy. [4]

Total Marks _____ / 21

Uses of Graphs & Other Graphs

1 A line is parallel to the line of equation $y = 3x - 2$ and goes through the point (1, 5).

Work out the equation of the line. [3]

2 Gemma, Naval and Esmai entered a five-mile cycling race. The graph below shows the race.

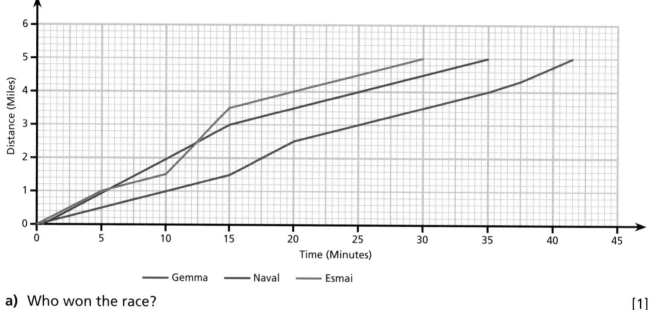

—— Gemma —— Naval —— Esmai

a) Who won the race? [1]

b) What speed was Naval travelling at for the last 20 minutes before he finished?
Give your answer in miles per hour. [2]

c) Between what times was Gemma travelling her fastest?
Give a reason for your answer. [2]

d) How many minutes after the race started did the winner move into the lead? [1]

e) Describe the race. [3]

3 The graph below shows the journey of a train. Work out the total distance travelled. [3]

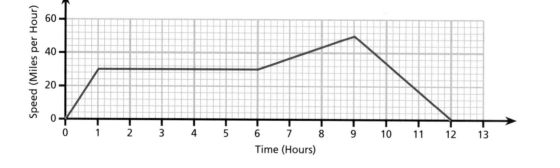

Total Marks .... / 15

Inequalities

1 Write down all the integer values for m that satisfy $-1 \leqslant m < 4$. [1]

2 Solve $2x - 6 > 2$ [2]

3 Write down all the possible integer values for y if $12 \leqslant 3y \leqslant 36$. [2]

4

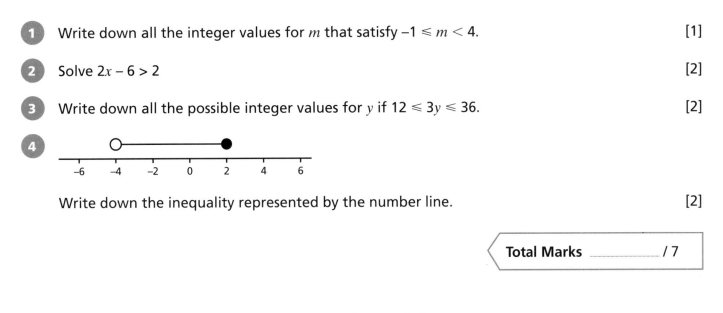

Write down the inequality represented by the number line. [2]

> **Total Marks** .... / 7

Congruence and Geometrical Problems

1 Prove that triangle ABC and triangle BCD are similar. [3]

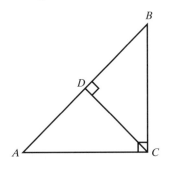

2 Lisa has a 10cm by 8cm photograph of her pet dog. She wants a smaller copy to fit into her handbag and a larger copy for her office.

a) What will the length of the smaller copy be, if the width is 4cm? [1]

b) What will the width of the larger copy be, if the length is 25cm? [2]

> **Total Marks** .... / 6

Right-Angled Triangles 1 & 2

1 A man walks 6.7km due north. He then turns due west and walks 7.6km.

How far is he now from his starting point? [3]

2 A 4m ladder leans against a vertical wire fence. The foot of the ladder is 2m from the base of the fence. Fang, the lion, can jump 3m vertically.

Will Fang be able to jump over the fence?
You must give reasons for your answer. [4]

3 ABC is an isosceles triangle. $AB = BC = 13$cm. D is the midpoint of AC and $AC = 10$cm.

Calculate the length of BD. [3]

4 A bumblebee leaves its nest and flies 10 metres due south and then 6 metres due west.

What is the shortest distance the bumblebee has to fly to return to its nest?
Give your answer to 3 significant figures. [3]

5 A triangle has side lengths of 1.5cm, 2.5cm and 2cm. Is it a right-angled triangle?
Give a reason for your answer. [3]

6 How long is the diagonal of a square of side length 3cm? [2]

7 A is the point (4, 0) and B is the point (7, 5).

Calculate the angle between line AB and the x-axis to the nearest degree. [2]

8 Molly cycles 5km in a north-easterly direction from Apton (A) to Bray (B). She then cycles 8km in a north-westerly direction from Bray to Chart (C).

a) How far is Chart from Apton? Give your answer to 2 significant figures. [2]

b) Calculate angle CAB. Give your answer to the nearest degree. [2]

9

Calculate:

a) The length of GH.
   Give your answer to 2 decimal places. [2]

b) The size of angle HGJ.
   Give your answer to 1 decimal place. [2]

Total Marks _____ / 28

Statistics 1 & 2

1 The table shows the number of pieces of fruit eaten by a group of students in one week.

Number of Pieces of Fruit	Frequency
10	6
11	
12	
13	10
14	10

a) Use the bar chart to complete the frequency table. [2]

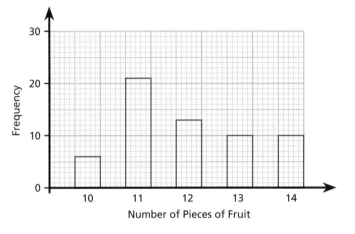

b) How many students were in the group? [1]

c) Write down the mode number of pieces of fruit eaten. [1]

d) Calculate the mean number of pieces of fruit eaten. [3]

2 The table below shows hours of sunshine and amount of rainfall for nine towns across England in one month.

Sunshine (h)	600	420	520	630	470	380	560	430	450
Rainfall (mm)	11	18	13	9	16	25	14	20	19

a) Draw a scatter diagram to represent this information. [2]

b) Describe the relationship between hours of sunshine and rainfall. [1]

Total Marks _____ / 10

Measures, Accuracy and Finance

You must be able to:

- Understand and solve problems relating to household finance
- Check calculations using approximation
- Round numbers and measures to an appropriate degree of accuracy
- Convert between metric units and use conversion factors to convert between imperial and metric units.

Solving Real-Life Problems

- Being able to understand and solve problems relating to finance is an essential life skill.

Nadine bought a designer handbag from an online shop for £230 and sold it two years later for £184.

Calculate the percentage loss.

Percentage profit or loss = $\dfrac{\text{profit or loss}}{\text{original amount}} \times 100$

$= \dfrac{£46}{£230} \times 100 = 20\%$ ← Loss = £230 – £184 = £46

VAT is charged at the standard rate of 20%.

What is the final cost after VAT has been added to a bill of £45?

$\dfrac{20}{100} \times £45 = £9$

Total bill = £45 + £9 = £54 ← This could also be calculated using a multiplier: 1.2 × £45 = £54

Yvonne earns £36 000 per year. The first £8200 is tax free, but income tax of 20% must be paid on the rest.

How much income tax does Yvonne have to pay each year?

£36 000 – £8200 = £27 800 ← Work out the amount that is taxable.

Income tax = $\dfrac{20}{100} \times £27\,800 = £5560$

Limits of Accuracy

- You will often be asked to give answers to a certain number of **decimal places** or **significant figures**. To do this, you must 'round off' the number.
- **Decimal places** refers to the number of digits after the decimal point, e.g.

27.3652 = 27.4 (to 1 d.p.) ← Because the digit two places after the decimal point is 5 or more, the 3 (in the first decimal place) rounds up to 4.

27.3652 = 27.37 (to 2 d.p.) ← The 5 rounds up the 6 to a 7.

- The first **significant figure** is the first non-zero figure (working from left to right), e.g.
  - 541 = 540 (to 2 significant figures)
  - 0.0347 = 0.035 (to 2 significant figures).
- Inequality notation is used to specify simple error intervals due to rounding, e.g. $144.5 \leqslant x < 145.5$

Approximation of Calculations

- You can use **approximations** to estimate and check the answers to calculations.

> Estimate the answer to $(2136 + 39.7) \div (9.6 \times 11.1)$
>
> $(2000 + 40) \div (10 \times 10)$
> $= 2040 \div 100 = 20.4$

Approximate each number to 1 significant figure.

Converting Between Units

LEARN

$$mm \xrightarrow{\div 10} cm \xrightarrow{\div 100} m \xrightarrow{\div 1000} km$$

$$mm \xleftarrow{\times 10} cm \xleftarrow{\times 100} m \xleftarrow{\times 1000} km$$

> How many metres are there in 4 kilometres?
>
> 4km = 4 × 1000 = 4000m

> A rod is 620 millimetres long. What is its length in centimetres?
>
> 620mm = 620 ÷ 10 = 62cm

- Imperial / metric conversions will be given in exam questions, but you must know how to use them.

> How many kilometres are there in 20 miles?
> Assume 8km = 5 miles
>
> 20 miles = 32km

Multiply by 4.

> How many pounds are there in 20 kilograms?
> Assume 1kg = 2.2 pounds
>
> 20kg = 44 pounds

Multiply by 20.

> **Quick Test** 📱
>
> 1. Estimate the value of $(0.897)^2 \times 392.4$
> 2. A lorry travels 150 miles between two towns.
>   If 5 miles is 8km, work out the distance travelled in
>   a) kilometres and b) metres.

Quadratic and Simultaneous Equations

You must be able to:

- Solve quadratic equations by factorising
- Solve two simultaneous equations
- Find approximate solutions to quadratic and simultaneous equations by using a graph.

Factorisation

- When solving a quadratic equation by **factorisation** (see p.15–17), make sure it equals zero first.

Solve the equation $x^2 + 4x + 3 = 0$ by factorisation.

×	x	$+1$
x	x^2	$+x$
$+3$	$+3x$	$+3$

$(x + 1)(x + 3) = 0$

$x + 1 = 0 \qquad x + 3 = 0$

$x = -1 \qquad x = -3$

> Set up and complete a table. The missing terms need to have a product of +3 and a sum of +4.

> First row = first bracket; first column = second bracket

The Method of Intersection

- Plotting a graph of a quadratic equation can give zero, one or two solutions for x when $y = 0$.
- The solutions are given by the curve's points of **intersection** with the x-axis.

Find approximate solutions to the equation $x^2 - 5x + 1 = 0$ by plotting a graph.

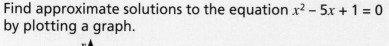

There are two solutions: $x = 0.2$ or $x = 4.8$

> These solutions are approximate.

Key Point

If two brackets have a product of zero, one of the brackets must equal 0.

Key Point

The points of intersection with the x-axis are called roots.

Simultaneous Equations

- **Simultaneous equations** can be solved by elimination.

Solve the following simultaneous equations:

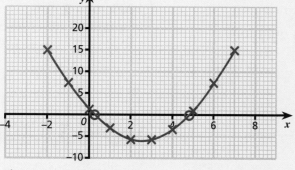

$3x - y = 18 \qquad$ Equation 1

$x + y = 10 \qquad$ Equation 2

$4x = 28 \qquad 7 + y = 10$

$x = 7 \qquad y = 3$

Key Point

Solutions to simultaneous equations always come in pairs.

> Add equation 1 and equation 2 to eliminate the y terms.

> Substitute your value for x into one of the equations.

Annabel buys three pears and two apples for £1.20.
David buys four pears and three apples for £1.65.

Work out the cost of one apple and one pear.

$$3p + 2a = 120 \quad \text{Equation 1}$$
$$4p + 3a = 165 \quad \text{Equation 2}$$

Equation 1 × 4: $12p + 8a = 480$

Equation 2 × 3: $12p + 9a = 495$

$$a = 15$$
$$3p + (2 \times 15) = 120$$
$$p = 30$$

An apple costs 15p and a pear costs 30p.

Form two equations with the information given.

Multiply so that the p terms match. Remember to multiply all terms.

Subtract equation 1 from equation 2.

Substitute your value for a into one of the equations and solve.

Solve the following equations simultaneously:

$y = 2x + 1 \quad$ Equation 1
$3y + x = 10 \quad$ Equation 2

$$3(2x + 1) + x = 10$$
$$6x + 3 + x = 10$$
$$7x + 3 = 10$$
$$7x = 7$$
$$x = 1$$
$$y = (2 \times 1) + 1$$
$$y = 3$$

Substitute $y = 2x + 1$ into equation 2.

Substitute your value for x into equation 1 to find y.

Solve the simultaneous equations:

$y = 3x^2$
$y + 5x = 3$

You can plot graphs and find the points of intersection. However the solutions are only approximate.

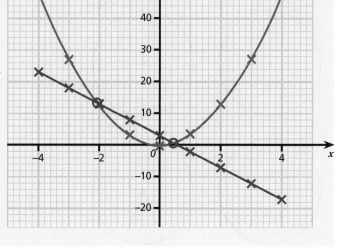

The points of intersection are
$(-2.1, 13.5)$ and $(0.5, 1)$.

So, the two approximate solutions are
$x = -2.1$ and $y = 13.5$ or $x = 0.5$ and $y = 1$.

Quick Test

1. Solve the equation $x^2 = 2x + 5$ by the method of intersection.
2. Solve the simultaneous equations:
   $2x + y = 5$
   $x + y = 3$

Key Words

factorisation
intersection
simultaneous equation

Circles

You must be able to:

- Identify the parts of a circle and understand their basic properties
- Recognise cyclic quadrilaterals
- Calculate angles in a circle.

Parts of a Circle

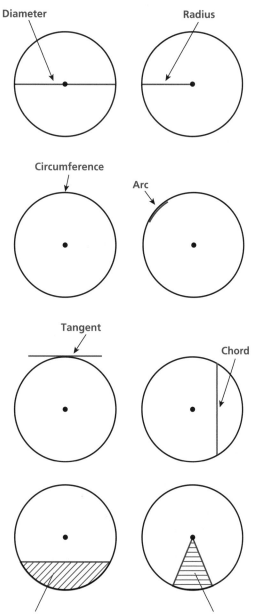

- The circumference is found with the formula:

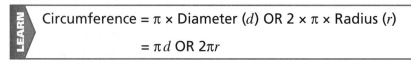

LEARN

Circumference = π × Diameter (d) OR 2 × π × Radius (r)

$$= \pi d \text{ OR } 2\pi r$$

Cyclic Quadrilaterals

- All four vertices (corners) of a **cyclic quadrilateral** touch the circumference of the circle.
- The **opposite angles** of a cyclic quadrilateral **add up to 180°.**

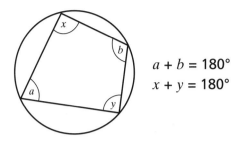

$a + b = 180°$

$x + y = 180°$

Angles in a Circle

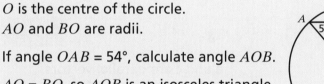

O is the centre of the circle.

AO and BO are radii.

If angle $AOB = 80°$, calculate angle ABO.

$AO = BO$, so AOB is an isosceles triangle.

Angle $ABO = \dfrac{(180° - 80°)}{2} = 50°$

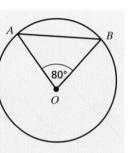

O is the centre of the circle.

AO and BO are radii.

If angle $OAB = 54°$, calculate angle AOB.

$AO = BO$, so AOB is an isosceles triangle.

Angle $OAB = OBA$

Angle $AOB = 180° - (54° + 54°) = 72°$

O is the centre of the circle.

If angle $EOF = 93°$, and angle $EOG = 132°$, calculate angle FOG.

Angle $FOG = 360° - (132° + 93°) = 135°$

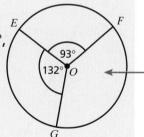

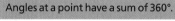 Angles at a point have a sum of 360°.

Key Words

diameter
radius / radii
circumference
arc
tangent
chord
segment
sector
cyclic quadrilateral

Quick Test

1. O is the centre of a circle. OX and OY are radii.
   If angle $OYX = 40°$, calculate angle XOY.
2. Draw a circle with a radius of 4cm.

Vectors

You must be able to:

- Add and subtract vectors
- Multiply a vector by a scalar
- Work out the magnitude of a vector
- Carry out translations according to column vectors.

Properties of Vectors

- A **vector** is a quantity that has both **magnitude** (size) and direction.
- Vectors are equal only when they have equal magnitudes and are in the same direction.

> **Key Point**
>
> The direction of a vector is shown by an arrow.

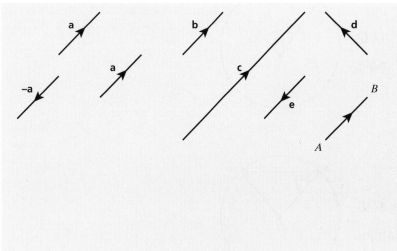

b = **a** (same direction, same length)

d ≠ **a** (different direction, same length)

e = –**a** (opposite direction, same length)

c = 3**a** (same direction, 3 × length of **a**)

$$\mathbf{a} = \overrightarrow{AB} = \underline{a} = \begin{pmatrix} 2 \\ 2 \end{pmatrix}$$

These are all ways of writing the same vector.

$$-\mathbf{a} = \begin{pmatrix} -2 \\ -2 \end{pmatrix}$$

a and **c** are parallel vectors.
a and **b** are equal vectors.

- Any number of vectors can be added together.

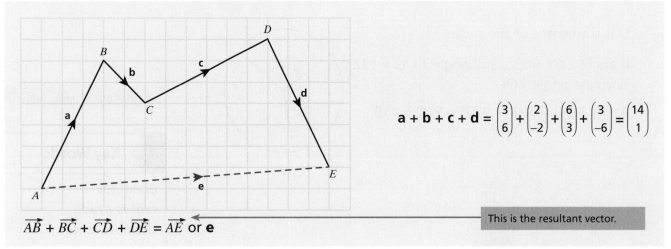

$$\mathbf{a} + \mathbf{b} + \mathbf{c} + \mathbf{d} = \begin{pmatrix} 3 \\ 6 \end{pmatrix} + \begin{pmatrix} 2 \\ -2 \end{pmatrix} + \begin{pmatrix} 6 \\ 3 \end{pmatrix} + \begin{pmatrix} 3 \\ -6 \end{pmatrix} = \begin{pmatrix} 14 \\ 1 \end{pmatrix}$$

$\overrightarrow{AB} + \overrightarrow{BC} + \overrightarrow{CD} + \overrightarrow{DE} = \overrightarrow{AE}$ or **e**

This is the resultant vector.

- When a vector is multiplied by a **scalar** (a numerical value), the resultant vector will always be parallel to the original vector.
- When a vector is multiplied by a positive number (not 1), the direction of the vector does not change, only its magnitude.

> **Key Point**
>
> The sum of the lengths $AB + BC + CD + DE$ does **not** equal the length of AE.

- When a vector is multiplied by a negative number (not –1), the magnitude of the vector changes and the vector points in the opposite direction.
- The magnitude of a vector **a** is written |**a**|
- Magnitude of vector $\begin{pmatrix} x \\ y \end{pmatrix}$ is $\sqrt{x^2 + y^2}$

Work out the magnitude of vector **a**.

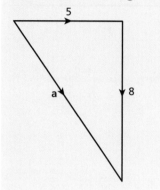

Vector $\mathbf{a} = \begin{pmatrix} 5 \\ -8 \end{pmatrix}$

$\mathbf{a}^2 = 5^2 + 8^2$ ← Use Pythagoras' Theorem.

$\mathbf{a}^2 = 25 + 64$

$|\mathbf{a}| = \sqrt{89}$

$|\mathbf{a}| = 9.43$ (to 3 significant figures)

Translations

- When a shape is translated, it does not change size or rotate. It moves left or right and up or down.
- The translation is represented by a column vector $\begin{pmatrix} x \\ y \end{pmatrix}$
- x represents the distance moved **horizontally**: **positive** means to the **right**, **negative** means to the **left**.
- y represents the distance moved **vertically**: **positive** means **up**, **negative** means **down**.

Translate the shaded shape by vector $\begin{pmatrix} -4 \\ -3 \end{pmatrix}$

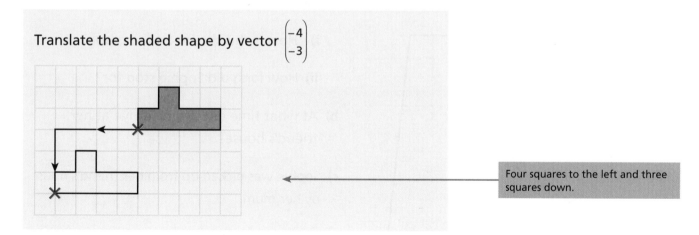

Four squares to the left and three squares down.

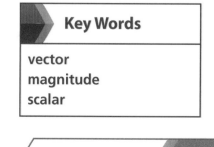

Key Words

vector
magnitude
scalar

Uses of Graphs & Other Graphs

1 The graph below shows the journey of a car.

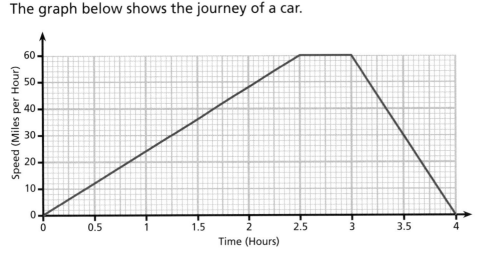

a) What is the greatest speed at which the car travels during the journey? [1]

b) Why does the graph have a gradient of zero between 2.5 and 3 hours? [1]

c) Calculate the total distance travelled by the car. [3]

2 The graph shows Sophie's journey to her friend's house. Her friend lives 18km away.
Sophie began her journey at 1pm.

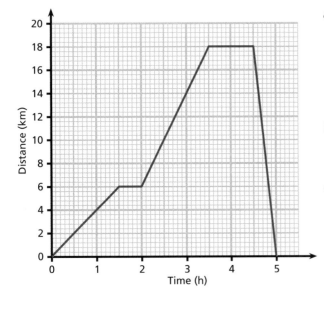

a) Sophie stopped on the way to see her friend.

   i) How is this shown on the graph? [1]

   ii) How long did Sophie stop for? [1]

b) At what time did Sophie arrive at her friend's house? [1]

c) Sophie was picked up from her friend's house by her mum.

   i) Calculate the average speed, in km/h, of her journey home. [2]

   ii) At what time did Sophie arrive home? [1]

3 Work out the equation of the line that is parallel to the line $y = 2x - 5$ and goes through the point (3, 6). [3]

Total Marks _____ / 14

Inequalities

1 Write down all the possible integer values of n if $3 \leqslant n \leqslant 7$. [1]

2 Solve $3x + 4 > 25$ [2]

3 Work out all the possible integer values of y if $15 \leqslant 5y \leqslant 35$. [2]

4 A TV salesperson is set a target to sell more than six televisions a week.
The manufacturer can let the salesperson have a maximum of 20 televisions each week.

Use an inequality equation to represent the number of televisions that could be sold
each week if the salesperson meets or exceeds their target. [2]

5 Maisie is thinking of a number (m). $11 < m < 17$
m is also a prime number.

What number is Maisie thinking of? [1]

Total Marks .... / 8

Congruence and Geometrical Problems

1 A triangle has angles of 56°, 64° and 60°. The triangle is enlarged by scale factor 2.

What are the angles of the enlarged triangle? [3]

2 A tree of height 5m casts a shadow that is 8.5m in length.

Work out the height of a tree casting a shadow that is 34m in length. [2]

3 What is the difference between congruency and similarity? [2]

4 Fill in the missing criteria to complete the sentence:

Two triangles are congruent if they satisfy one of four criteria: SSS, RHS, SAS and .... [1]

Total Marks .... / 8

Review Questions

Right-Angled Triangles 1 & 2

1   A rectangle has a length of 10cm and a width of 5cm.

Calculate the length of its diagonal. Give your answer to 3 significant figures.   [2]

2   A square has a diagonal length of 12cm.

Calculate the side length of the square. Give your answer to 2 decimal places.   [3]

3 PQR is a triangle. P is at (1, 0), Q is at (1, 5) and R is at (6, 0).

   a) What type of triangle is PQR?   [1]

   b) Calculate the length of PR.   [1]

   c) Calculate the length of PQ.   [1]

   d) Calculate the length of QR. Give your answer to 2 decimal places.   [3]

   e) Work out the area of triangle PQR.   [2]

4   Sean and Alexander are arguing about a triangle that has side lengths of 9cm, 40cm and 41cm. Sean says it is a right-angled triangle and Alexander says it is not.

Who is correct? Write down a calculation to support your answer.   [3]

5   Moira is standing 80m from the base of Blackpool Tower. The angle of elevation to the top of the tower is 63.15°.

Calculate the height of the tower to the nearest metre.   [3]

6   The coordinates of a triangle XYZ are X(1, 4), Y(1, 1) and Z(6, 1).

Calculate angle XZY. Give your answer to 3 decimal places.   [2]

7   Chevaun says that sin 30° + sin 60° > cos 30° + cos 60°.

Is she correct? Show working to support your answer.   [3]

8   A helicopter leaves an air base in London and flies 175km in a north-easterly direction to an air base in Norwich.

   a) How far north of the London air base is the Norwich air base?
      Give your answer to the nearest kilometre.   [3]

   b) How long did it take the helicopter to make the journey if it travelled 100km in 30 minutes?   [2]

Total Marks .... / 29

Statistics 1 & 2

 1 The table shows information about the genre of the books sold by a bookshop one Friday.

Book Genre	Frequency
Non-Fiction	20
Crime	15
Children's	50
Science Fiction	5

    a) How many books were sold in total? [1]

    b) Draw a pie chart to represent this information. [3]

2 Corinna wants to sample 100 people in Malmesbury to find out how often people visit the library. She stands by the entrance to the library and asks the first 100 people she meets.

    a) Suggest a reason why this is not a representative sample. [1]

    b) Describe a better method that Corinna could use to collect her data. [2]

3 The scatter graph below shows the body temperature and pulse rate of ten animals.

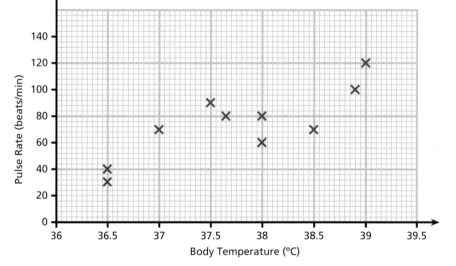

    a) Draw a line of best fit and use it to estimate the pulse rate in an animal with a body temperature of 39.5°C. [2]

    b) Comment on the reliability of this estimate. [1]

Total Marks _____ / 10

Practice Questions

Measures, Accuracy and Finance

1 Estimate an answer to the following calculation: 🖩

$$\frac{512 \times 7.89}{38.9}$$

[2]

2 Four kilograms of carrots and five kilograms of potatoes cost £3.76.

If five kilograms of potatoes cost £1.80, work out the cost of one kilogram of carrots. [2]

3 Sabrina's take-home pay is worked out using this formula:

Take-Home Pay = Hours Worked × Hourly Rate – Deductions

Sabrina's hourly rate is £35.
Her deductions were £218.
Her take-home pay was £657.

Work out the number of hours she worked that week. [3]

4 Calculate $\dfrac{\sqrt{(6.2^2 - 3.6)}}{2.6 \times 0.15}$

Give your answer to:

a) 2 decimal places [3]

b) 3 significant figures. [1]

5 Write down the most appropriate metric unit for measuring:

a) The distance between Portsmouth and Eastbourne. [1]

b) The quantity of flour needed for a small cake. [1]

c) The capacity of a can of lemonade. [1]

6 A marathon is about 25 miles long. If 5 miles = 8km, how many kilometres is this? [1]

7 Round 8765 to:

a) The nearest thousand [1]

b) The nearest hundred [1]

c) The nearest ten. [1]

Total Marks .... / 18

Quadratic and Simultaneous Equations

1 Solve the equation $3x^2 = 27$

Circle your answer.

$x = 3$ $\qquad\qquad$ $x = -3$ $\qquad\qquad$ $x = \dfrac{\pm\sqrt{27}}{3}$ $\qquad\qquad$ $x = \pm 3$ [1]

2 **a)** On the same set of axes, draw the graphs $y = 2x^2$ and $y = 3x + 2$ [2]

$\quad$ **b)** Use your graph to solve the equation $2x^2 = 3x + 2$ [2]

3 Solve the simultaneous equations: [4]

$2x + y = 1$

$y = x - 2$

4 **a)** Factorise $x^2 + 2x - 8$ [1]

$\quad$ **b)** Use your answer to part **a)** to help solve the equation $x^2 + 2x - 8 = 0$ [2]

5 Rebecca goes to the greengrocer's shop and buys three apples and two pears. She pays £2.20.

Mandeep goes to the same greengrocer's shop and buys six apples and two pears. She pays £3.40.

Work out the cost of one pear and the cost of one apple.
You **must** show all your working. [4]

6 Solve the simultaneous equations:

$x + y = 5$

$2x - y = 7$ [4]

7 Solve the simultaneous equations:

$4y + 3x = 17$

$3y + 2x = 12$ [4]

Practice Questions

8 The graph has the equation $y = x^2 + 2x - 2$

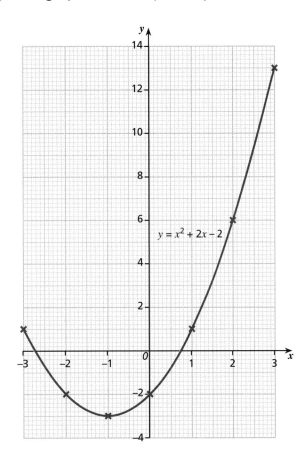

$y = x^2 + 2x - 2$

a) Using the graph, estimate solutions to the equation $x^2 + 2x - 2 = 0$ [2]

b) On the same axes, draw the graph of equation $y = x - 1$ [1]

c) Use your graphs to solve the equation $x^2 + 2x - 2 = x - 1$ [2]

9 Solve the simultaneous equations:

$4x + y = 17$

$2x + y = 9$ [4]

10 Solve the following equation by factorising:

$x^2 + 3x = 4$ [3]

11 Circle the correct value of a for:

$x^2 + 7x + 6 = (x + a)(x + 1)$

$a = 6$ $\qquad a = 8$ $\qquad a = 7$ $\qquad a = 3$ [1]

Total Marks .... / 37

Circles

1   For each of the following questions, work out the lettered angles.
The centre of each circle is marked with an O where appropriate. 🖩   [6]

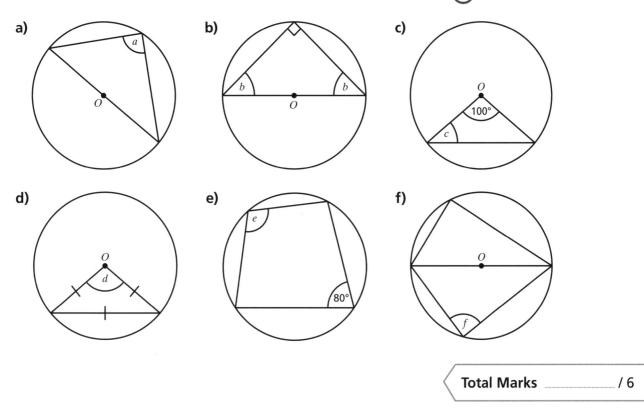

a)   b)   c)

d)   e)   f)

Vectors

1   On squared paper, draw a set of axes that go from –7 to +7 in each direction.

a) Plot the points A(1, 2), B(5, 2) and C(5, 0) and join them together. Label the shape E.

What shape have you drawn?   [2]

b) Translate shape E by the vector $\begin{pmatrix} 0 \\ -4 \end{pmatrix}$. Label the shape F.

What are the coordinates of shape F?   [3]

c) Translate shape E by the vector $\begin{pmatrix} -7 \\ 3 \end{pmatrix}$. Label the shape G.

What are the coordinates of shape G?   [3]

Review Questions

Measures, Accuracy and Finance

1 Calculate the value of $\dfrac{25.75 \times 31.3}{7.62 - 1.48}$

Give your answer to **a)** 2 decimal places and **b)** 3 significant figures. [3]

2 The Cotton family, comprising of Mum, Dad and 16-year-old daughter, are going away to Turkey.

The brochure states that for every three days booked, you receive one day free.
The price per adult per day is £42. There is a 15% reduction for a child up to the age of 17.
The package is all inclusive.

a) What would the total cost of the holiday be if the family books 12 days? [3]

b) There is an additional early booking discount of 5%.

   If the Cotton family book early, how much do they save? [2]

c) The flight to Turkey takes 4 hours 20 minutes. The departure time from Gatwick was scheduled for 11.10am but, due to bad weather, the flight was delayed by 1 hour 34 minutes.

   If Turkish time is two hours ahead of UK time, at what time did the flight arrive in Turkey? [2]

d) The distance from Gatwick to Dalyan (in Turkey) is 3672km.

   If 5 miles = 8km, what is the distance from Gatwick to Dalyan in miles? [2]

3 Complete the sentences:

a) The mass of an egg is about _____ grams. [1]

b) The floor to ceiling height of a room is about _____ metres. [1]

c) The volume of an average size mug is about _____ millilitres. [1]

> Total Marks _____ / 15

Quadratic and Simultaneous Equations

1 The area of the triangle is 1.5cm². [4]

Work out the value of x.

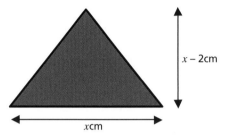

$x - 2$cm

$x$cm

2 The graph below shows the line with equation $y = x^2 - 5x + 6$

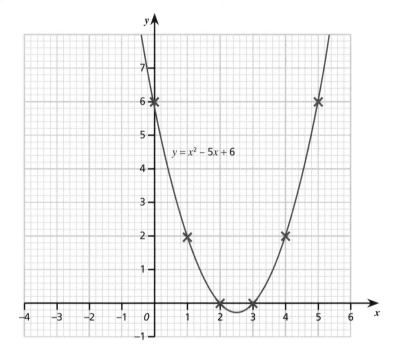

$y = x^2 - 5x + 6$

a) On the same axes, draw the line of the equation $y = 2$. [1]

b) Use the graphs to solve the equation $x^2 - 5x + 6 = 2$ [2]

3 Solve $x^2 + 8x - 9 = 0$ [3]

4 Stephen buys and downloads three apps and four singles at a total cost of £6.10.
Martin buys and downloads five apps and two singles at a total cost of £5.50.

Work out the cost of one app and one single. [4]

5 The sum of two numbers is 20 and the difference is 4.

Set up a pair of simultaneous equations and solve to find the two numbers. [4]

6 The graph has the equation $y = 3x^2$

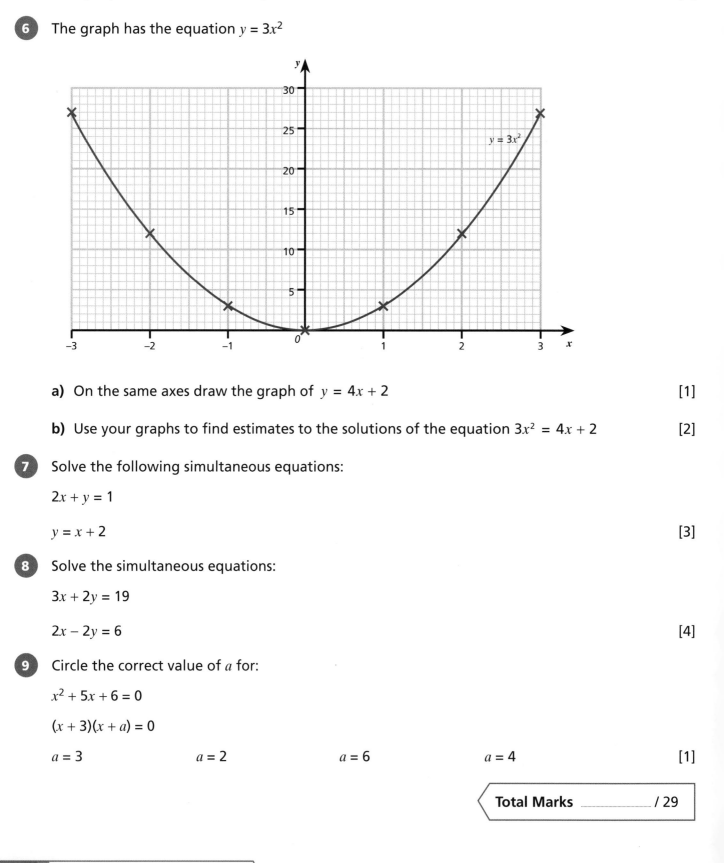

a) On the same axes draw the graph of $y = 4x + 2$ [1]

b) Use your graphs to find estimates to the solutions of the equation $3x^2 = 4x + 2$ [2]

7 Solve the following simultaneous equations:

$2x + y = 1$

$y = x + 2$ [3]

8 Solve the simultaneous equations:

$3x + 2y = 19$

$2x - 2y = 6$ [4]

9 Circle the correct value of a for:

$x^2 + 5x + 6 = 0$

$(x + 3)(x + a) = 0$

$a = 3$ $\qquad$ $a = 2$ $\qquad$ $a = 6$ $\qquad$ $a = 4$ [1]

Total Marks .... / 29

Circles & Vectors

1 Complete the sentences:

    a) A curved line that is part of the circumference of a circle is an _____. [1]

    b) A straight line from the centre of the circle to the circumference is a _____. [1]

    c) The distance around the edge of the circle is the _____. [1]

    d) A line through the centre of the circle, with both ends touching the

        circumference, is the _____. [1]

    e) A line that touches the circumference at both ends, but does not pass through the

        centre, is a _____. [1]

    f) A line outside the circle, which touches the circumference of the circle at only

        one point, is a _____. [1]

    g) _____ is the number 3.142 (to 3 decimal places). [1]

    h) The formula used to calculate the area of a circle is _____. [1]

    i) The formula used to calculate the circumference of a circle is _____. [1]

2 Which of these vectors will be parallel to $4\mathbf{a} + 3\mathbf{b}$? Circle your answer.

    $4\mathbf{a} + 6\mathbf{b}$ $8\mathbf{a} + 3\mathbf{b}$ $8\mathbf{a} + 6\mathbf{b}$ $4\mathbf{a} - 3\mathbf{b}$ [1]

3 Translate shape A by the vector $\begin{pmatrix} -8 \\ 1 \end{pmatrix}$ [1]

Total Marks _____ / 11

Mixed Exam-Style Questions

1 The diagram is a regular pentagon of side length P.

Shade the area on the diagram that is represented by the expression $\frac{1}{2}CG + \frac{1}{2}PG$.

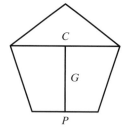

[1]

2 Factorise completely $3x^3 + 6x^2$

Circle your answer.

$3(x^3 + 2x^2)$ $x(3x^2 + 6x)$ $x^2(3x + 6)$ $3x^2(x + 2)$ [1]

3 Solve $3(x + 3) - 5 = 2(x - 2)$

Answer _____ [3]

4 Work out the value of $\frac{5}{7}\left(x - \frac{2}{5}\right) + 9y$ when $x = \frac{1}{5}$ and $y = \frac{2}{7}$

Answer _____ [2]

5 Simplify $3x^2 + 6x - 2x^2 + 4x - 2y - 6x^2$

Answer _____ [2]

6 Which of the expressions below is equivalent to $3p^2y - py + p^2y + 7py$?
Circle your answer.

$10p^2y$ $\qquad$ $4p^2y - 8py$ $\qquad$ $4p^2y + 6py$ $\qquad$ $2p^2y + 6py$ $\qquad$ [1]

7 Solve $5(x - 3) = -3$

Answer _____ [2]

8 Factorise $x^2 - x - 2$

Answer _____ [1]

9 Use the formula $P = 3r - q^2$ to work out the value of q when $P = 30$ and $r = 20$.

Give your answer to 2 decimal places.

Answer _____ [2]

10 **a)** Factorise $x^2 - 16$

Answer _____ [2]

b) Use your answer to part **a)** to solve $x^2 - 16 = 0$

Answer _____ [2]

11 The formula used to calculate the area of a circle is $A = \pi r^2$.

A circle has an area of 25cm².

Ethan thinks the radius of the circle is $\dfrac{5}{\sqrt{\pi}}$

Guy thinks the radius is $\dfrac{\sqrt{\pi}}{5}$

Who is correct? Write down a calculation to support your answer.

Answer _____ [2]

12 Circle the largest number.

6.77   6.767   6.677   6.8   [1]

13 **a)** Write 45 as a product of prime factors.

Answer _____ [2]

b) Write 105 as a product of prime factors.

Answer _____ [1]

c) Use your answers to parts **a)** and **b)** to work out the highest common factor of 45 and 105.

Answer _____ [2]

14 Work out $5\frac{1}{6} - 2\frac{1}{3}$ 🖩

Answer _____ [3]

15 $P = xy$

x is increased by 10%.
y is increased by 10%.

Work out the percentage increase in P.

Answer _____ [2]

16 Mandeep is looking for a new 12-month phone contract.

Dave's Dongles	Ian's Internet
£12 a month	£10 a month
+	+
5p a minute	6p a minute
10% discount on first 6 months	15% discount on first 4 months

On average Mandeep uses 120 minutes per month.

Which phone contract is cheaper for Mandeep? 📵
You must show your working.

Answer _____ [5]

17 **a)** Circle the calculation that is equivalent to 97×1452 📵

9.7×14520 9.7×145.2 970×14520 0.97×145.2 [1]

b) $97 \times 1452 = 140844$ 📵

Use this information to write down the value of 9.7×145.2

Answer _____ [1]

18 $y = \dfrac{ab}{a + b}$

$a = 3 \times 10^4$

$b = 5 \times 10^3$

Work out the value of y. Give your answer in standard form.

Answer _____ [2]

19 What is the formula for calculating speed (S)? Circle your answer.

$S = \dfrac{\text{Distance}}{\text{Time}}$ $\qquad$ $S = \text{Distance} \times \text{Time}$ $\qquad$ $S = \dfrac{\text{Distance}^2}{\text{Time}}$ $\qquad$ $S = \text{Time}^2 \times \text{Distance}$ $\quad$ [1]

20 A regular polygon has 20 sides. Circle the size of each of its interior angles.

18° $\qquad\qquad$ 162° $\qquad\qquad$ 3240° $\qquad\qquad$ 9° $\qquad\qquad$ [1]

21 Circle the correct expression for the nth term of the following sequence of numbers:

8, 11, 14, 17, 20 …

$3n + 5$ $\qquad\qquad$ $3 + 5n$ $\qquad\qquad$ $8n + 3$ $\qquad\qquad$ $3n + 8$ $\qquad$ [1]

22 A cat rescue centre recorded the age of the cats it re-homed over the course of a year.

Age	Frequency
1	10
2	21
3	16
4	12
5	6

a) Write down the modal age.

Answer _____ [1]

b) Calculate the mean age. Give your answer to 2 decimal places.

Answer _____ [3]

c) On the graph paper below draw a bar chart to represent the data.

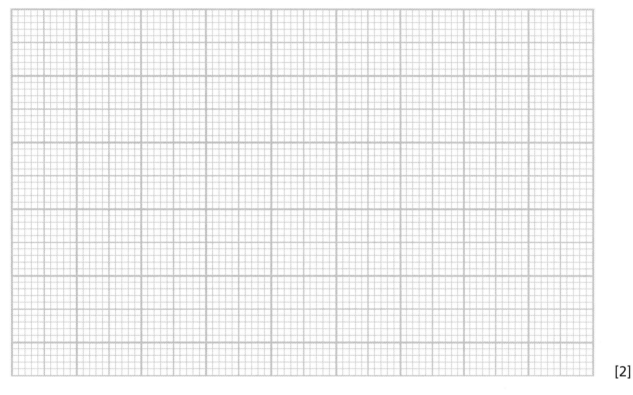

[2]

23 The force (F) between two objects is directly proportional to the distance (x) between them.

$F = 4$ when $x = 2$

a) Work out an expression for F in terms of x.

Answer _____ [2]

b) Work out the value of F when $x = 5$.

Answer _____ [1]

c) Work out the value of x when $F = 20$.

Answer _____ [2]

24 Circle the equation for the y-axis.

$y = 0$ $x = 0$ $y = x$ $y = -x$ [1]

25 As part of a health and safety review, a company surveys its employees to find out how many wear glasses or contact lenses.

	Male	Female
Glasses	9	6
Contact Lenses	8	16
Neither	20	15

a) Write down the ratio of the number of females who wear glasses to the number of females who wear contact lenses. Give your answer in its simplest form.

Answer _____ [1]

b) What is the percentage of all employees who are male and do not wear glasses or contact lenses?

Answer _____ [2]

26 Sketch the graph of $y = x^2 + 5x + 4$ [1]

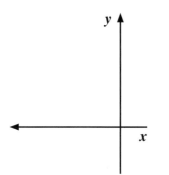

27 Solve the simultaneous equations:

$y = 2x - 1$

$y = 3x - 4$

Answer _____ [3]

28 In the diagram, A, B and C are points on the circumference of a circle, centre O.

Angle $BCE = 57°$

FE is a tangent to the circle at point C.

a) Calculate the size of angle ACB. Give a reason for your answer.

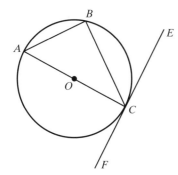

_____

_____ [2]

b) Calculate the size of angle BAC.
   Give reasons for your answer.

_____ [2]

29 Simplify $(5t^{-2})^{-1}$

Answer _____ [2]

30 The diagram shows a sector of a circle, centre O.
The radius of the circle is 6cm.
The angle of the sector at the centre of the circle is 115°.

Work out the perimeter of the sector.

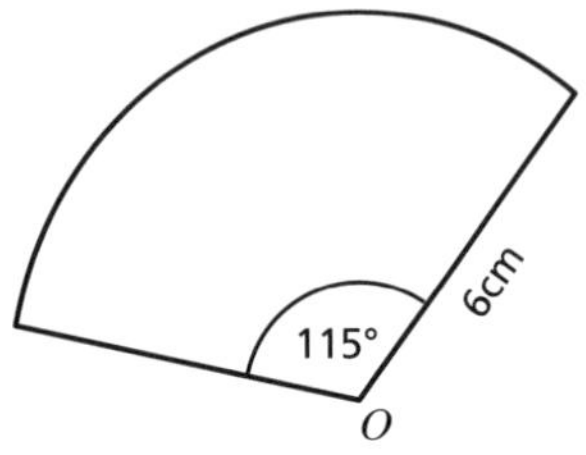

Answer _____ [4]

31 A bag contains six blue beads and five red beads.
Samuel takes a bead at random from the bag. He records its colour and replaces it.
He does this one more time.

Work out the probability that he takes one bead of each colour from the bag.

Answer _____ [3]

32 A sketch of the graph of the equation $y = x^2 + 3x + 2$ is shown below.

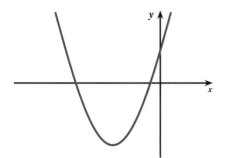

a) Circle the correct roots of the graph.

–3 and –1   –2 and –1   –4 and –2   –3 and –2   [1]

b) Work out the coordinates of the turning point.

Answer _____ [3]

Total Marks _____ / 80

Answers

Pages 6–7 Review Questions

1. 2 000 005   [1]
2. 45kg   [1]
3. $2^2 \times 3 \times 5$ [1]
4. a) 18   [1]
   b) 180   [1]
5. 8   [1]
6. −8   [1]
7. 64   [1]
8. $3 + 2m$ [1]
9. $3y^2$ [1]
10. a) 17, 20   [1]
    b) Yes [1]; nth term = $3n + 2$ and
    $(3 \times 46) + 2 = 140$ [1] OR
    $\frac{(140 - 5)}{3} = 45$
    (term-to-term rule is +3) [1]
11. 0.06   [1]
12. 9 sides   [1]
13. 3.46   [1]
14. (4.5, 7)   [2]
15. 27   [1]
16. 10km/h   [1]
17. 125cm³   [1]
18. 0.4   [1]
19. 0.875   [1]
20. $\frac{78}{100}$ [1]; $\frac{39}{50}$ [1]
21. $\frac{2}{3} = \frac{8}{12} = \frac{16}{24}$ and $\frac{3}{4} = \frac{9}{12} = \frac{18}{24}$ [1]; $\frac{17}{24}$ [1]
22. 10, 15, 20   [3]
23. 999 999   [1]
24. £21 : £35   [2]
25. 0.04, 0.39, 0.394, 0.4   [1]
26. a) $m = 4$ [1]
    b) $m = 48$ [1]
    c) $m = 1.5$ [1]
27. 36cm²   [1]
28. 404   [1]
29. £12.75   [1]
30. $13a - 10y$ [1]
31. £81   [1]
32. $A = \pi \times r \times r$ [1]; $A = 314$cm² [1]
33. Vol (1) = $2 \times 2 \times 2 = 8$cm³ [1];
    Vol (2) = $4 \times 4 \times 4 = 64$cm³ [1];
    $\frac{64}{8} = 8$ [1]
34. 15   [1]
35. $\frac{63}{3} \times 4$ [1]; = 84 [1]
36. $m = -3$ [1]
37. (5, 0)   [2]
38. 13cm   [1]
39. $(2 \times 3^2) + (3 \times 4) = 18 + 12$ [1];
    = 30 [1]
40. 6.5   [1]

Pages 8–35 Revise Questions

Page 9 Quick Test
1. 14 boxes
2. Molly. The multiplication must be carried out before the addition.

Page 11 Quick Test
1. a) 21
   b) −13
2. 1.63×10^{-3}

3. − 2201, − 220, 220, 1022, 2200
4. £12.16

Page 13 Quick Test
1. $2^2 \times 3^2$
2. a) 3
   b) 60
3. $m = 3$; $p = 11$
4. 10am

Page 15 Quick Test
1. $5(x + 2)$
2. $x = 2$
3. $6y - 5$
4. −171
5. $12t^2 - 3t$

Page 17 Quick Test
1. $w = 9$
2. $(x + 7)(x + 1)$
3. $q = \frac{2t + 5}{6}$
4. $y = \frac{x + 8}{6}$

Page 19 Quick Test
1. £108
2. 4 : 1
3. a) Box A, 1 sachet costs 31.7p; Box B, 1 sachet costs 31.1p; Box B is the best buy
   b) 2

Page 21 Quick Test
1. 402mph
2. £51.11
3. 50%

Page 23 Quick Test
1. a) Rectangle, Parallelogram, Kite
   b) Square, Rhombus
2. Angle $EJH = 16°$

> When FE and GH are made longer, they produce a triangle, FJH. Angles EHF and EFG are alternate angles on parallel lines.

Page 25 Quick Test
1. a) 3240°
   b) 162°
2. 12
3.

Bearing = 150° ± 2°

Page 27 Quick Test
1. $\frac{3}{4} \div \frac{1}{5}$ $\left(\frac{3}{4} + \frac{1}{5} = \frac{15}{20} + \frac{4}{20} = \frac{19}{20}\right.$ and
   $\left.\frac{3}{4} \div \frac{1}{5} = \frac{15}{4} = 3\frac{3}{4}\right)$
2. a) $0.\dot{1}$
   b) recurring
3. $\frac{3}{80}$

Page 29 Quick Test
1. a) $\frac{3}{20}$
   b) 0.15
2. 27 Turkish Lira
3. 32%
4. 15%, $\frac{1}{5}$, $\frac{5}{20}$, 1.15
5. James is correct.
6. Any 12 squares shaded.

Page 31 Quick Test
1. 275 carrots
2. £10.50 + £60 = £70.50
3. 1.5%

Page 33 Quick Test
1. a) 0.1
   b) 45
   c) Yes, the probabilities are not equal.

Page 35 Quick Test
1. $\frac{1}{18}$
2. $\frac{29}{81}$

Pages 36–41 Practice Questions

Page 36
1. a) 3.84×10^5 [2] (1 mark for correct digits but decimal point in wrong position)
   b) 384 000 − 19 600 [1]; = 364 400 or 3.644×10^5 [1]
2. 5°C   [1]
3. $(−3 − 6) − (−25)$ [2]
4. 84   [1]
5. 60   [1]
6. Fifth square number = 25 [1]; third cube number = 27 [1]; The third cube number is greater [1]
7. $256 = 2^8$ [1]; $n = 8$ [1]
8. No [1]; $2^3 = 8$, $3^2 = 9$ [1]
9. 2 packs of sausages [1]; 3 packs of bread rolls [1]

Page 37
1. $5x + 30$ [2] (1 mark for each correct term)
2. $5(3x + 2)$ [1]
3. $30 − 12x$ [2] (1 mark for each correct term)
4. $4t = p + q$ [1]; $t = \frac{p + q}{4}$ [1]
5. $9x + 4y$ [2] (1 mark for each correct term)
6. $4y − 8z + 12$ [2]
7. $4x^2 + 2x + 4$ [1]
8. 23   [1]
9. $8b = 14$ or $2b = \frac{7}{2}$ [1]; $b = \frac{7}{4}$ or $1\frac{3}{4}$ [1]
10. $3p + 6 = 2p + 6$ [1]; $p = 0$ [1]

11. $\frac{5}{2}x - \frac{2}{3}x = \frac{1}{2} + \frac{1}{3}$ **[1]**; $\frac{15}{6}x - \frac{4}{6}x = \frac{3}{6} + \frac{2}{6}$ **[1]**;

$x = \frac{5}{11}$ **[1]**

> Look for a common denominator.

12. $6x - 30y + 36$ **[2]** (1 mark for 2 correct terms)
13. $6p - 4q + 12$ **[2]**

> $- \times - = +$

14. $4xz(y - 1)$ **[2]** OR $4z(yx - x)$ OR $4x(yz - z)$ **[1]**
15. $(x + 1)(x + 2)$ **[2]** (1 mark for each correct bracket)
16. $6x - 15 + 4x + 12 - 4x$ **[1]**; $6x - 3$ **[1]**; $3(2x - 1)$ **[1]**
17. $A = \frac{1}{2}(a + b)h$ **[1]**

Page 38
1. $1 : 2000$ **[1]**
2. $180° \div 9 = 20°$ **[1]**; largest angle = $80°$ **[1]**
3. 4 people = 6 celery sticks, 1 person = $6 \div 4 = 1.5$ sticks (OR 3.5×6) **[1]**; 20 people = $20 \times 1.5 = 30$ sticks **[1]**
4. 6×4 **[1]**; = 24 days **[1]**

1. Density = mass ÷ volume **[1]**; Density = $4560 \div 400 = 11.4$g/cm³ **[1]**
2. Speed = distance ÷ time **[1]**; Speed = $200 \div 22 = 9.09$m/s **[1]**
3. Speed = distance ÷ time **[1]**; Speed = $15 \div 3 = 5$km/h **[1]**
4. Final amount = Original amount $\times \left(1 + \frac{\text{Rate}}{100}\right)^{\text{Time}}$ **[1]**; Final Amount = $4000 \times \left(1 + \frac{4}{100}\right)^4 = 4000 \times 1.04^4 = £4679.43$ **[1]**; Compound Interest = $£4679.43 - £4000 = £679.43$ **[1]**

Page 39
1.

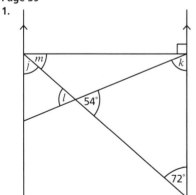

(Marks will not be awarded if reason is incorrect.) $j = 72°$ (alternate angle) **[1]**; $k = 54°$ (sum of the interior angles of a triangle = 180°) **[1]**; $l = 54°$ (vertically opposite angles are equal) **[1]**; $m = 18°$ ($90° - 72°$) **[1]**
2. $360° - 46° - 107° - 119° = 88°$ **[1]**
3. Exterior angle = $\frac{360}{n} = \frac{360}{10} = 36°$ **[1]**
4. Bearing = $180° + 36° = 216°$ **[1]**
5. 36km ÷ 4 = 9, 9cm **[1]**
6. $315°$ **[1]**

Page 40
1. $32 \div 8 = 4$ **[1]**; $32 - 4 = 28$ books **[1]**
2. $\frac{1}{3} + \frac{1}{6} + \frac{1}{4} = \frac{4}{12} + \frac{2}{12} + \frac{3}{12} = \frac{9}{12}$ **[1]**; Fraction that are horses is $1 - \frac{9}{12} = \frac{3}{12} = \frac{1}{4}$ **[1]**
3. $\frac{1}{6}, \frac{4}{12}, \frac{12}{24}, \frac{2}{3}$ **[2]**
4. a) $\frac{4}{5} \times \frac{2}{3} = \frac{8}{15}$m² **[1]**
   b) $\frac{4}{5} + \frac{4}{5} + \frac{2}{3} + \frac{2}{3} = \frac{44}{15}$m **[2]** (1 mark if answer is given as $2\frac{14}{15}$)
5. a) $\frac{45}{100}$ **[1]**; $\frac{9}{20}$ **[1]**
   b) $\frac{7.000}{8}$ **[1]**; = 0.875 **[1]**

1. $£4800 \times \frac{100}{80}$ **[1]**; = £6000 **[1]**
2. $\frac{9}{36} \times 100$ **[1]**; = 25% **[1]**
3. $\frac{28}{120} \times 100 = 23.33\%$ **[1]**

Page 41
1. a) Fully correct table **[2]** (1 mark for a list of 36 outcomes)

	1	2	3	4	5	6
1	1	2	3	4	5	6
2	2	4	6	8	10	12
3	3	6	9	12	15	18
4	4	8	12	16	20	24
5	5	10	15	20	25	30
6	6	12	18	24	30	36

   b) $\frac{13}{36}$ **[1]**
   c) $\frac{8}{36} = \frac{2}{9}$ **[1]**
2. a) $\frac{1}{2}$ **[1]**
   b) $\frac{1}{2}$ **[1]**
3. a) $\frac{305}{500} = \frac{61}{100}$ **[1]**
   b) $\frac{97}{500} \times 100 = 19.4$ **[1]**; 19 or 20 **[1]**
4. a) i) $\frac{6}{10}$ OR $\frac{3}{5}$ **[1]**
      ii) $\frac{3}{10}$ **[1]**
      iii) $\frac{1}{10}$ **[1]**
   b) $\left(\frac{6}{10} \times \frac{6}{10}\right) + \left(\frac{3}{10} \times \frac{3}{10}\right) + \left(\frac{1}{10} \times \frac{1}{10}\right)$ **[1]**; $\frac{23}{50}$ **[1]**
5. a)

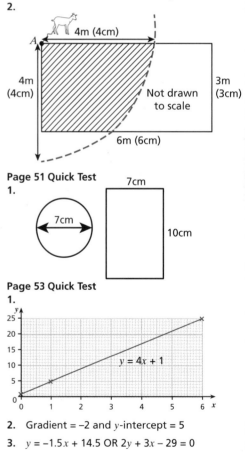

   15 **[1]**   10   5 **[1]**   5 **[1]**

   b) 5 **[1]**

Pages 42–63 Revise Questions

Page 43 Quick Test
1. $-4, -8$
2. $-2, 43$
3. 72

Page 45 Quick Test
1. 18, 21
2. a) $3n + 4$
   b) 154
3. $n^2 - 1$

Page 47 Quick Test
1. a) A translation by the vector $\begin{pmatrix} 5 \\ -1 \end{pmatrix}$
   b) A reflection in the line $y = x$
   c) A rotation of 90° anticlockwise about (0, 0)

Page 49 Quick Test
1. Construct an angle of 60°and bisect it, i.e. draw a line and mark on it two points, A and B.

   Open compasses to length AB.

   Put compass point on A and draw an arc. Put compass point on B and draw an arc.

   Draw a line to join A to the new point, C.

   Adjust compasses so less than length AB.

   Put compass point on A and draw arcs crossing AB and AC at points D and E.

   Put compass point on D and draw an arc. Put compass point on E and draw an arc. Draw a line from A to the new point, F.

2.

Not drawn to scale

Page 51 Quick Test
1.

7cm

7cm   10cm

Page 53 Quick Test
1.

$y = 4x + 1$

2. Gradient = -2 and y-intercept = 5
3. $y = -1.5x + 14.5$ OR $2y + 3x - 29 = 0$

Answers

Page 55 Quick Test

1.

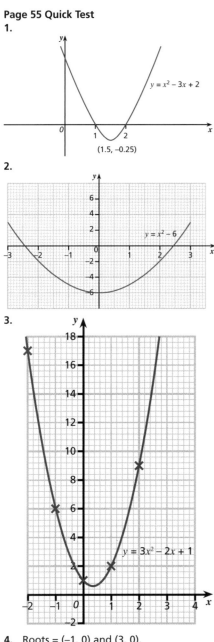

$y = x^2 - 3x + 2$

(1.5, −0.25)

2.

$y = x^2 - 6$

3.

$y = 3x^2 - 2x + 1$

4. Roots = (−1, 0) and (3, 0),
turning point = (1, −3)

Page 57 Quick Test

1. $6x^5$
2. 2^8
3. $\frac{1}{27}$
4. $\frac{x^4}{y^{10}}$

Page 59 Quick Test

1. Area = 12m², perimeter = 14m
2. 10cm²
3. Volume = 18cm³, surface area = 42cm²

Page 61 Quick Test

1. Volume = 301.59cm³ (to 2 d.p.) and surface area = 251.33cm² (to 2 d.p.) OR 301.44cm³ and 251.2cm² if using π = 3.14
2. 24cm²

---

3. Circumference = 21.99cm (to 2 d.p.) and area = 38.48cm² (to 2 d.p.) OR 21.98cm and 38.47cm² if using π = 3.14

Page 63 Quick Test

1. $\frac{500\pi}{3}$ or 523.6cm³
2. 91.5cm²
3. $\frac{200}{3}$ or 66.7cm³
4. 108π or 339.3cm²

Pages 64–69 Review Questions

Page 64
1. 8 adult tickets = £308.80 and 4 child tickets = £51.04 **[1]**; £359.84 **[1]**
2. a) 311.961 **[1]**
   b) 0.311961 **[1]**
3. −27 **[1]**
4. 24 combinations **[1]**
5. a) 9 **[1]**
   b) 35 **[1]**
   c) 9 OR 100 **[1]**
   d) 27 **[1]**
   e) 13 **[1]**
6. 64 **[1]**
7. Peter is correct **[1]**; $(2^3)^2 = 2^6 = 64$, $(2^2)^3 = 2^6 = 64$ **[1]**
8. $3 \times 5 \times 7$ **[2]**

Page 65
1. $k - g$ **[1]**
2. $12x - 5y$ **[2]** (1 mark for each correct term)
3. $\frac{7}{9}$ **[1]**
4. $6 + 10 = 5x - 2x$ **[1]**; $16 = 3x$ **[1]**; $x = \frac{16}{3}$ **[1]**
5. $b(5a - 3bc)$ **[1]**
6. $4x - 12 = 10$ **[1]**; $4x = 22$ **[1]**; $x = 5.5$ **[1]**
7. $x = -6$ **[1]**
8. a) $18x + 14$ **[1]**
   b) $18x + 14 = 56$ **[1]**; $x = \frac{7}{3} = 2\frac{1}{3}$ cm **[1]**
9. a) $r^2 = \frac{V}{\pi h}$ **[1]**; $r = \sqrt{\frac{V}{\pi h}}$ **[1]**
   b) $r = \sqrt{\frac{50}{\pi \times 10}}$ **[1]**; $r = 1.26$ (to 3 significant figures) **[1]**

Page 66
1. $10^2 = k \times 5$ **[1]**; $100 = 5k$, $k = 20$ **[1]**
2. 1 part = £60 ÷ 12 = £5 **[1]**; Shares are (5 × £5) = £25, (7 × £5) = £35 **[1]**; Difference = £10 **[1]**
3. 6 hours = 360 minutes **[1]**; 360 : 4 = 90 : 1 **[1]**

1. a) Distance = Speed × Time = 1.5 × 1.5 **[1]**; 2.25m **[1]**
   b) $\frac{64}{8} \times 5 = 40$ miles **[1]**;
   Zebra runs at 40mph, so the antelope is faster **[1]**
2. Mass = Density × Volume = 2700 × 4.5 **[1]**; = 12150kg **[1]**
3. Pressure = $\frac{\text{Force}}{\text{Area}}$ **[1]**; $\frac{588}{0.0001}$ = 5880000Pa **[1]**

Page 67
1. $y + 2y + 3y = 180°$, $6y = 180°$, $y = 30°$ **[1]**; Largest angle = 90° **[1]**
2. 80° + 123° + 165° + 40° = 408° It is not quadrilateral, because the interior angles are greater than 360° in total **[1]**
3. Kite, rectangle, parallelogram, arrowhead **[2]** (1 mark for two or three correct)
4. Bearing = 180° + 054° = 234° **[1]**
5. 60° **[1]**
6. a) Sum = 6 × 180° **[1]**; 1080° **[1]**
   b) $\frac{1080°}{8} = 135°$ **[1]**
7. 3 × 6.5 = 19.5km **[1]**
8. a) $y + 5 + 3y - 16 + 2y + 5 = 180°$, $6y - 6 = 180°$ **[1]**
   b) $6y = 186°$, $y = 31°$ **[1]**
   c) $y + 5 = 36°$ **[1]**; $3y - 16 = 77°$ **[1]**; $2y + 5 = 67°$ **[1]**
9. Pauline is correct **[1]**; exterior angle is 180° − 158° = 22° **[1]**; number of sides = 360° ÷ 22 = 16.36̇; it cannot be right because the number of sides must be a whole number **[1]**

Page 68
1. $6 \times \frac{4}{3} = 8$ **[1]**
2. $\frac{1}{16}$ **[1]**
3. $\frac{1}{8} \times 344 = 43$ **[1]**; 344 − 43 = 301 biscuits **[1]**
4. $\frac{(95 \times 6)}{5}$ **[1]**; £114 **[1]**
5. $\frac{(5.000)}{8} = 0.625$ **[1]**

1. $\frac{35}{100} \times £56 = £19.60$ **[1]**; £56 − £19.60 = £36.40 **[1]**
2. Richard is correct **[1]**; $\frac{30}{100} \times £40 = £12$, $\frac{40}{100} \times £30 = £12$ **[1]**
3. 39.5% **[1]**
4. $\frac{15}{500} \times 100 = 3\%$ **[1]**

Page 69
1. a) 0.3 **[1]**
   b) 50 × 0.3 **[1]**; = 15 **[1]**
   c) No **[1]**; P(Georgia wins) = 0.6, so greater than 0.5 (0.5 would be fair). **[1]**
2. 28 + 10 + 35 = 73 **[1]**; 82 − 73 **[1]**; 9 **[1]**
3. a) Venn diagram with no crossover **[1]**; All labels correct **[1]**

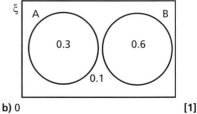

   b) 0 **[1]**
   c) 0.3 + 0.6 **[1]**; = 0.9 **[1]**

> Remember, mutually exclusive means P(A and B) = 0

---

4. $\frac{6}{10}$ **[1]**; $\frac{3}{5}$ **[1]**

Pages 70–75 Practice Questions

Page 70

1. a) i) 20, 23 **[1]**
   ii) +3 **[1]**
   b) i) −4, −8 **[1]**
   ii) −4 **[1]**
   c) i) $\frac{7}{9}, \frac{7}{27}$ **[1]**
   ii) ÷3 **[1]**
2. 24 − 4n **[1]**
3. −2, 1, 4, 7, 10 **[2]** (1 mark for any three correct)
4. a) 3n + 11 **[2]** (1 mark for each correct term)
   b) 311 **[1]**
5. a) 23 **[1]**; 30 **[1]**
   b) 3 **[1]**
6. a)

   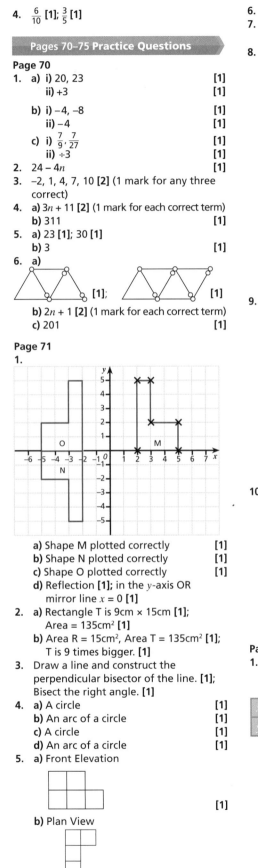

   [1]; **[1]**
   b) 2n + 1 **[2]** (1 mark for each correct term)
   c) 201 **[1]**

Page 71

1.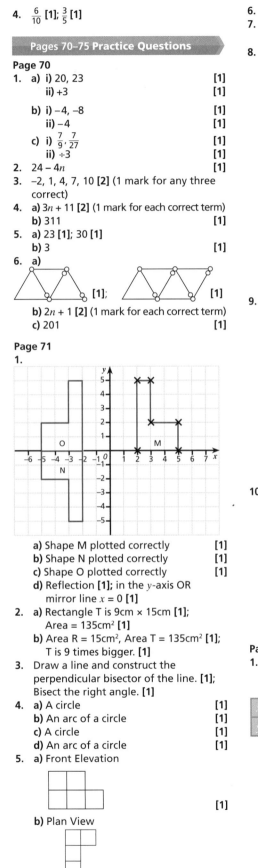

   a) Shape M plotted correctly **[1]**
   b) Shape N plotted correctly **[1]**
   c) Shape O plotted correctly **[1]**
   d) Reflection **[1]**; in the y-axis OR mirror line x = 0 **[1]**
2. a) Rectangle T is 9cm × 15cm **[1]**; Area = 135cm² **[1]**
   b) Area R = 15cm², Area T = 135cm² **[1]**; T is 9 times bigger. **[1]**
3. Draw a line and construct the perpendicular bisector of the line. **[1]**; Bisect the right angle. **[1]**
4. a) A circle **[1]**
   b) An arc of a circle **[1]**
   c) A circle **[1]**
   d) An arc of a circle **[1]**
5. a) Front Elevation

   [1]

   b) Plan View

   [1]

---

6. A, B and C **[1]**
7. a) 2 **[1]**
   b) 2 **[1]**
8.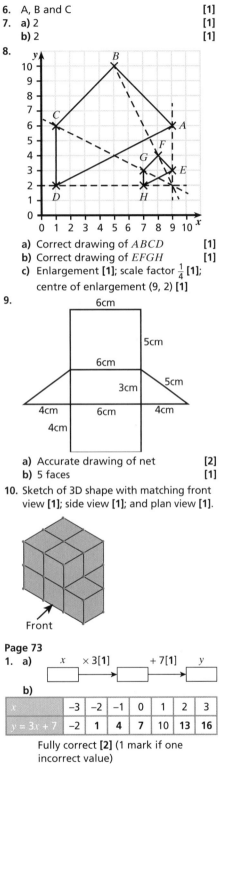

   a) Correct drawing of $ABCD$ **[1]**
   b) Correct drawing of $EFGH$ **[1]**
   c) Enlargement **[1]**; scale factor $\frac{1}{4}$ **[1]**; centre of enlargement (9, 2) **[1]**
9.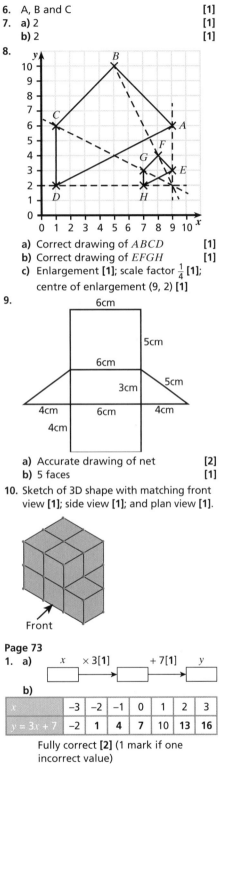

   a) Accurate drawing of net **[2]**
   b) 5 faces **[1]**
10. Sketch of 3D shape with matching front view **[1]**; side view **[1]**; and plan view **[1]**.

    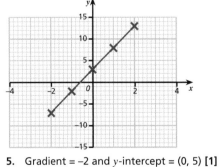

    Front

Page 73

1. a) x ─→ ×3 **[1]** ─→ +7 **[1]** ─→ y
   b)

x	−3	−2	−1	0	1	2	3
$y = 3x + 7$	−2	1	4	7	10	13	16

   Fully correct **[2]** (1 mark if one incorrect value)

---

c) Correctly plotted points **[1]**; straight line drawn **[1]**

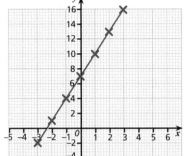

2. a) Sketch showing correct shaped curve **[1]**; correct x-intercepts **[1]**; and correct y-intercept. **[1]**

   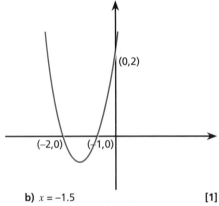

   (0,2)

   (−2,0) (−1,0)

   b) $x = -1.5$ **[1]**
3. Fully correct graph with y-intercept at (0, −2) **[1]**; and a straight line crossing through points (−4, −18) and (4,14) **[1]**

   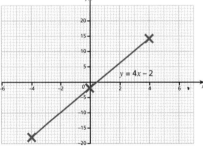

   $y = 4x - 2$

4. Fully correct graph with y-intercept at (0, 3) **[1]**; and a straight line crossing through points (−2, −7) and (2,13) **[1]**

5. Gradient = −2 and y-intercept = (0, 5) **[1]**

Answers

6. Fully correct table **[1]**; and accurately plotted graph **[1]**

x	–3	–2	–1	0	1	2	3
y	34	17	6	1	2	9	22

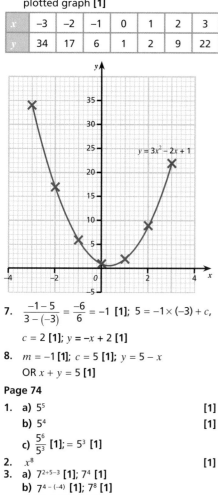

$y = 3x^2 - 2x + 1$

7. $\frac{-1-5}{3-(-3)} = \frac{-6}{6} = -1$ **[1]**; $5 = -1 \times (-3) + c$,

$c = 2$ **[1]**; $y = -x + 2$ **[1]**

8. $m = -1$ **[1]**; $c = 5$ **[1]**; $y = 5 - x$
OR $x + y = 5$ **[1]**

Page 74

1. a) 5^5 **[1]**
b) 5^4 **[1]**
c) $\frac{5^6}{5^3}$ **[1]**; $= 5^3$ **[1]**

2. x^8 **[1]**

3. a) 7^{2+5-3} **[1]**; 7^4 **[1]**
b) $7^{4-(-4)}$ **[1]**; 7^8 **[1]**
c) $7^{3\times-2}$ **[1]**; 7^{-6} **[1]**

4. Rebecca **[1]**; $2^2 \times 5^2 = 4 \times 25 = 100$ **[1]**

Because 2 and 5 are different numbers, the basic rules of indices do not apply.

5. 9^{-2} OR $x^{2\times-2}$ **[1]**; $\frac{1}{81x^4}$ OR $\frac{1}{81}x^{-4}$ **[1]**

6. $x^{-1} + 1$ **[2]** (1 mark for each correct term)

7. $27r^6p^3$ **[2]**

8. a) 8×4 **[1]**; 32 **[1]**
b) $\frac{27}{2}$ **[1]**; 13.5 **[1]**
c) $\frac{1}{64}$ **[1]**
d) 1 **[1]**

Page 75

1. a) Perimeter =
$7 + (2 \times 1) + (2 \times 5) + 3 + (2 \times 2)$
(or equivalent) **[1]**; 26cm **[1]**
b) Area = $(7 \times 1) + (5 \times 3)$ **[1]**; 22cm^2 **[1]**

2. $6 \times 10 \times x = 300$ **[1]**; $x = 5\text{cm}$ **[1]**

3. Area of square = 36cm^2 **[1]**; Area of circles = $2 \times \pi \times 1.5^2 = 14.137...$ **[1]**; Shaded region = $36 - 14.137...$ **[1]**; $= 21.9\text{cm}^2$ (to 3 significant figures) **[1]**

4. a) Volume of large cylinder =
$\frac{3}{4} \times 6000\pi = 4500\pi$ **[1]**;
$4500\pi = \pi \times 15^2 \times h$ **[1]**;
$h = 20\text{cm}$ **[1]**

b) Volume of small cylinder =
$\frac{1}{4} \times 6000\pi = 1500\pi$ **[1]**;
$1500\pi = \pi \times r^2 \times h = \pi \times r^3$ **[1]**;
$r = 11.4\text{cm}$ (to 3 significant figures) **[1]**

5. Volume of cone = $\frac{1}{3} \times \pi \times 3^2 \times 7 = 21\pi$ **[1]**;
Volume of the hemisphere =
$\frac{1}{2} \times \frac{4}{3} \times \pi \times 3^3 = 18\pi$ **[1]**; Volume of plastic needed = $21\pi + 18\pi = 39\pi \text{ cm}^3$ **[1]**

Pages 76–91 Revise Questions

Page 77 Quick Test
1. $y = -2x + 15$
2. 2

Page 79 Quick Test
1.

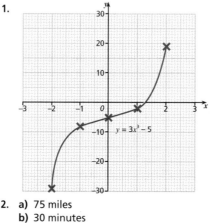

$y = 3x^3 - 5$

2. a) 75 miles
b) 30 minutes

Page 81 Quick Test
1.

$x < 7$

2. $-5, -4, -3, -2$
3. a) $>$
b) $<$

Page 83 Quick Test
1. a) Angle ABC = Angle ADE (corresponding angles); Angle ACB = Angle AED (corresponding angles); Angle DAE is common to both triangles; so triangles are similar (three matching angles).
b) $\frac{5}{10} = \frac{BC}{8}$ so $BC = 4\text{cm}$

Page 85 Quick Test
1. 13cm
2. $8^2 + 15^2 = 17^2$
3. $24^2 + 56^2 = C^2$, $C = 60.9\text{cm}$

Page 87 Quick Test
1. $\sin\theta = \frac{3.5}{8} = 0.4375$, $\theta = 25.90°$
2. $\cos 35° = \frac{x}{14}$, $x = 11\text{km}$

Page 89 Quick Test
1. a)

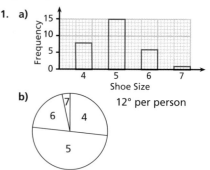

b) 12° per person

c) The pie chart, to show the proportion of pupils in the class who wear the different shoe sizes.

Page 91 Quick Test
1. a) 7.76
b) 8
c) 9
d) 3

Pages 92–97 Review Questions

Page 92
1. LCM of 3 and 4 = 12 **[1]**; therefore, 29 will be common to both **[1]**
2. a)

Pattern Number	Perimeter (cm)
1	5
2	8
3	11
4	14
60	182 **[1]**
n	$3n + 2$ **[1]**

b) $3n + 2 < 1500$ **[1]**; 499 pentagons **[1]**
3. $-5, -2, 3$ **[2]** (1 mark for one or two correct terms)
4. 66 **[1]**; 130 **[1]**

The difference is doubled each time.

Page 93–94
1. a) $X = (-5, 1)$ **[1]**; $Y = (-3, 5)$ **[1]**; $Z = (-1, 2)$ **[1]**
b) $X = (1, -5)$ **[1]**; $Y = (5, -3)$ **[1]**; $Z = (2, -1)$ **[1]**
2. Lengths of rectangle D: $3 \times 3 = 9\text{cm}$ and $5 \times 3 = 15\text{cm}$ **[1]**; Area of rectangle C = $3 \times 5 = 15\text{cm}^2$, area of rectangle D = $9 \times 15 = 135\text{cm}^2$ **[1]**; Ratio = 15 : 135 = 1 : 9 **[1]**
3. Volume of cuboid C = $3 \times 4 \times 5 = 60\text{cm}^3$ **[1]**; Volume of cuboid D = $9 \times 12 \times 15 = 1620\text{cm}^3$ **[1]**; Ratio = 60 : 1620 = 1 : 27 **[1]**
4. $A' = (3, 2)$ **[1]**; $C' = (6, 6)$ **[1]**; $B' = (5, 3)$ **[1]**; $D' = (4, 5)$ **[1]**
5. a) A vertical line **[1]**
b) An arc of a circle **[1]**
c) A horizontal straight line **[1]**
d) An arc of a circle **[1]**

6. A square 4cm × 4cm **[1]**

7.

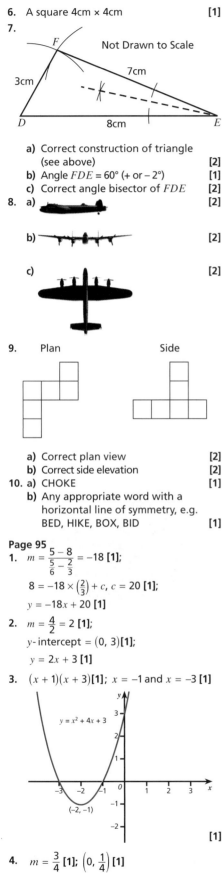

Not Drawn to Scale

a) Correct construction of triangle
   (see above) **[2]**
b) Angle FDE = 60° (+ or − 2°) **[1]**
c) Correct angle bisector of FDE **[2]**

8. a) **[2]**

b) **[2]**

c) **[2]**

9. Plan   Side

a) Correct plan view **[2]**
b) Correct side elevation **[2]**

10. a) CHOKE **[1]**
b) Any appropriate word with a
   horizontal line of symmetry, e.g.
   BED, HIKE, BOX, BID **[1]**

Page 95
1. $m = \dfrac{5-8}{\frac{5}{6}-\frac{2}{3}} = -18$ **[1]**;

$8 = -18 \times \left(\frac{2}{3}\right) + c, c = 20$ **[1]**;

$y = -18x + 20$ **[1]**

2. $m = \frac{4}{2} = 2$ **[1]**;
y-intercept = (0, 3)**[1]**;
$y = 2x + 3$ **[1]**

3. $(x + 1)(x + 3)$**[1]**; $x = -1$ and $x = -3$ **[1]**

$y = x^2 + 4x + 3$

(−2, −1)

[1]

4. $m = \frac{3}{4}$ **[1]**; $\left(0, \frac{1}{4}\right)$ **[1]**

5. a) $y = (x + 7)(x - 1)$ **[1]**; $x^2 + 6x - 7$ **[1]**;
   $a = 6, b = -7$ **[1]**

b) −3 **[1]**

6. a)

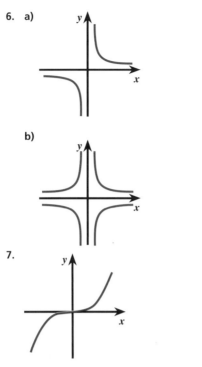

[1]

b)

[1]

7.

[1]

Page 96
1. $\frac{1}{8}a^{-3}b^{15}$ **[1]**; $9a^{-4}b^6$ **[1]**; $\frac{9}{8}a^{-7}b^{21}$ **[1]**

2. True **[1]**; $x^{-6} = \frac{1}{x^6}$ **[1]**

3. a) 3^6 **[1]**
   b) 3^3 **[1]**
   c) 3^6 **[1]**

4. a) 5^{4+5-2} **[1]**; 5^7 **[1]**
   b) $5^{5-(-4)}$ **[1]**; 5^9 **[1]**
   c) $5^{3\times-6}$ **[1]**; 5^{-18} **[1]**

5. Darcey **[1]**; $3^2 + 3^3 = 9 + 27 = 36$ **[1]**

> The basic rules of indices do not
> apply when adding terms.

6. 16^{-2} OR $x^{3\times-2}$ **[1]**; $\frac{1}{256x^6}$ OR $\frac{1}{256}x^{-6}$ **[1]**

7. $x^2 + x^8$ **[2]** (1 mark for each correct term)

Page 97

1. a) $\frac{1}{2} \times 6 \times 8 \times 9$ **[1]**; 216cm³ **[1]**

   b) $\sqrt[3]{216}$ **[1]**; 6 × 12 **[1]**; = 72cm **[1]**

2. $\pi r^2 = 2\pi r$ **[1]**; $r = 2$ **[1]**

3. $\frac{1}{2}(2x + x) \times 3x \times 20 = 900$ **[1]**;
   $9x^2 = 90$ **[1]**; $x^2 = 10$ **[1]**;
   $x = \sqrt{10}$ or 3.16cm **[1]**

4. $75 = 4 \times \pi \times r^2$ **[1]**;

   $r^2 = \frac{75}{4\pi}$ **[1]**; $r = 2.44$cm (to 3 significant

   figures) **[1]**

5. $\frac{1}{2} \times x \times 6 = 7.5$ **[1]**; $6x = 15$ **[1]**;
   $x = 2.5$ **[1]**

6. 5 × 3.5 = 17.5m² **[1]**; 0.5 × 5 × 1 = 2.5m²
   [1]; 17.5 + 2.5 = 20m² **[1]**; 2 tins **[1]**

Pages 98–101 Practice Questions

Page 98
1. $m = 3$ **[1]**; $5 = 3 + c, c = 2$ **[1]**; $y = 3x + 2$ **[1]**

2. a) Esmai **[1]**
b) 6mph **[2]** (1 mark for 2 miles in
   20 minutes)
c) 15–20 minutes **[1]**; the part of
   Gemma's graph with the steepest
   gradient **[1]**
d) 12.5 minutes **[1]**
e) Esmai and Naval are level for first 5
   minutes **[1]**; Esmai then speeds up
   and overtakes Naval **[1]**; Esmai wins,
   Naval is second and Gemma is third. **[1]**
   (Accept any other valid points.)

3. $\frac{1}{2} \times 1 \times 30 + 5 \times 30 + \frac{1}{2} \times 3 \times 50$**[1]**;

   $+\frac{1}{2} \times (30 + 50) \times 3$ **[1]**; 360 miles **[1]**

> Break the area under the graph
> down into two triangles, a rectangle
> and a trapezium.

Page 99
1. −1, 0, 1, 2, 3 **[1]**
2. $2x > 8$ **[1]**; $x > 4$ **[1]**
3. $4 \leqslant y \leqslant 12$ **[1]**; 4, 5, 6, 7, 8, 9, 10,
   11, 12 **[1]**
4. $-4 < x \leqslant 2$ **[2]** (1 mark for each side of
   inequality)

1. Angle ACB = Angle BDC = 90° **[1]**;
   Angle ABC = Angle DBC (the angle is
   common to both triangles) **[1]**; Angle
   BAC = Angle BCD (180° − Angle
   B − 90°), so the triangles are similar
   (three matching angles) **[1]**

2. a) 5cm **[1]**
   b) 20cm **[2]**

Page 100
1. $7.6^2 + 6.7^2 = y^2$ **[1]**; $57.76 + 44.89 = y^2$ **[1]**;
   $y = \sqrt{(102.65)} = 10.13$km **[1]**
2. $4^2 - 2^2 = f^2$ **[1]**; $16 - 4 = f^2$ **[1]**; $\sqrt{12} = f$,
   fence height = 3.46m **[1]**; No, because
   the fence height is 3.46m and Fang can
   only jump 3m. **[1]**
3. $13^2 - 5^2 = BD^2$ **[1]**; $169 - 25 = BD^2$,
   $BD = \sqrt{144}$ **[1]**; BD = 12cm **[1]**
4. $10^2 + 6^2 = b^2$ **[1]**; $100 + 36 = b^2$ **[1]**;
   $b = \sqrt{136} = 11.7$m (to 3 significant
   figures) **[1]**
5. $1.5^2 + 2^2 = 6.25$ **[1]**; $\sqrt{6.25} = 2.5$ **[1]**; The
   triangle is right-angled (Pythagoras'
   Theorem) **[1]**.
6. $3^2 + 3^2 = d^2$ **[1]**; $d = \sqrt{18} = 4.24$cm **[1]**
7. $\tan\theta = \frac{5}{3} = 1.6667$ **[1]**; $\theta = 59°$ **[1]**
8. a) $5^2 + 8^2 = CA^2$, $89 = CA^2$ **[1]**; $CA = \sqrt{89}$
   = 9.4km **[1]**
   b) $\sin CAB = \frac{8}{9.4} = 0.851$ **[1]**; $CAB = 58°$
   [1] OR $\tan CAB = \frac{8}{5} = 1.6$ **[1]**;
   $CAB = 58°$ **[1]**

> Making a sketch will show that you
> are dealing with a right-angled
> triangle.

Answers

9. a) $\sin 32° = \dfrac{GH}{9}$ **[1]**;
$GH = \sin 32° \times 9 = 4.77$cm **[1]**

b) $\tan HGJ = \dfrac{6}{4.77}$ **[1]**;
$\tan HGJ = 1.2579$, $51.5°$ **[1]**

Page 101

1. a)

Number of Pieces of Fruit	Frequency
10	6
11	21 **[1]**
12	13 **[1]**
13	10
14	10

b) 60 **[1]**

c) 11 **[1]**

d) $6 \times 10 + 21 \times 11 + 13 \times 12 + 10 \times 13 + 10 \times 14$ **[1]**; $717 \div 60$ **[1]**; 11.95 pieces of fruit **[1]**

2. a) Drawn and labelled axes **[1]**; and accurately plotted points. **[1]**

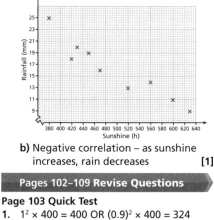

b) Negative correlation – as sunshine increases, rain decreases **[1]**

Pages 102–109 Revise Questions

Page 103 Quick Test

1. $1^2 \times 400 = 400$ OR $(0.9)^2 \times 400 = 324$

2. a) 240km
b) 240000m

Page 105 Quick Test

1. $x = $ (approx.) -1.4 or 3.4
2. $x = 2$, $y = 1$

Page 107 Quick Test

1. Angle $XOY = 100°$
2. Circle of radius 4cm

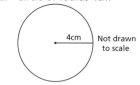

4cm   Not drawn to scale

Page 109 Quick Test

1. a) N

120°
3cm

b) N

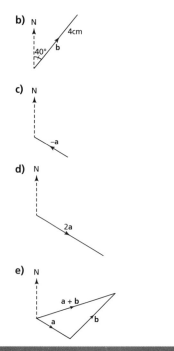

4cm
40° b

c) N

$-a$

d) N

2a

e) N

a + b
a   b

Pages 110–113 Review Questions

Page 110

1. a) 60mph **[1]**
b) Constant speed **[1]**
c) $0.5 \times 2.5 \times 60 + 0.5 \times 60 + 0.5 \times 1 \times 60$ **[1]**; $75 + 30 + 30$ **[1]**; 135 miles **[1]**

> Break the area under the graph down into two triangles and a rectangle.

2. a) i) A straight horizontal line **[1]**
ii) 30 minutes **[1]**
b) 4.30pm **[1]**
c) i) $\dfrac{18}{0.5}$ **[1]**; 36km/h **[1]**
ii) 6pm **[1]**
3. Gradient $= 2$ **[1]**; $y = 2x + c$, $6 = (2 \times 3) + c$ **[1]**; $y = 2x$ **[1]**

Page 111

1. $n = 3, 4, 5, 6, 7$ **[1]**
2. $3x > 21$ **[1]**; $x > 7$ **[1]**
3. $3 \leqslant y \leqslant 7$ **[1]**; 3, 4, 5, 6, 7 **[1]**
4. If t is number of televisions, $t > 6$ and $t \leqslant 20$ **[1]**; $6 < t \leqslant 20$ **[1]**
5. 13 **[1]**

1. 56° **[1]**; 64° **[1]**; 60° **[1]**
2. Shadow is $\dfrac{34}{8.5}$ times bigger $= 4$ times bigger **[1]**; height of tree $= 5 \times 4 = 20$m **[1]**
3. Similar figures are identical in shape but can differ in size **[1]**; congruent figures are identical in shape and size. **[1]**
4. AAS **[1]**

Page 112

1. $10^2 + 5^2 = d^2$ **[1]**; $d = \sqrt{125}$, $d = 11.2$cm **[1]**
2. $x^2 + x^2 = 12^2$ **[1]**; $2x^2 = 144$, $x^2 = 72$ **[1]**; $x = \sqrt{72} = 8.49$cm **[1]**
3. a) Right-angled triangle or an isosceles triangle **[1]**

b) $PR = 5$ units **[1]**
c) $PQ = 5$ units **[1]**
d) $5^2 + 5^2 = QR^2$ **[1]**; $QR^2 = 50$, $QR = \sqrt{50}$ **[1]**; $QR = 7.07$ units **[1]**
e) Area $= \dfrac{1}{2}$ (base × height) **[1]**; $\dfrac{1}{2} \times 5 \times 5 = 12.5$ square units **[1]**

4. Sean was correct **[1]**; $9^2 = 81$, $40^2 = 1600$, $41^2 = 1681$ so $9^2 + 40^2 = 41^2$ **[1]** Pythagoras' Theorem works, so triangle is right-angled. **[1]**

5. $\tan \theta = \dfrac{\text{opp}}{\text{adj}}$, $\tan 63.15° = \dfrac{x}{80}$ **[1]**; $x = 80 \times \tan 63.15°$ **[1]**; $x = 158$m **[1]**

6. $\tan \theta = \dfrac{3}{5}$ **[1]**; $\tan \theta = 0.6$, $\theta = 30.964°$ **[1]**

7. $\sin 30° + \sin 60° = 0.5 + 0.8660 = 1.366$ **[1]**; $\cos 30° + \cos 60° = 0.8660 + 0.5 = 1.366$ **[1]**; Both answers are the same – Chevaun is incorrect **[1]**

8. a) $\sin 45° = \dfrac{x}{175}$ **[1]**; $x = 175 \times \sin 45° = 123.7$km **[1]**; $x = 124$km **[1]**
b) 100km in 30 minutes, 50km in 15 minutes, 25km in 7.5 minutes **[1]**; 175km in 52.5 minutes **[1]**

Page 113

1. a) 90
b) 1 book $= \dfrac{360°}{90} = 4°$ **[1]**; non-fiction $= 80°$, crime $= 60°$, children's $= 200°$, science fiction $= 20°$ **[1]**; accurately drawn pie chart **[1]**

SF 20°
Non Fiction 80°
Children's 200°
Crime 60°

2. a) Biased, as she will only speak to people who use the library **[1]**
b) A random **[1]**; selection of 100 people who live in Malmesbury **[1]**

3. a) Line of best fit drawn as per diagram below **[1]**; Accept 120 to 135 **[1]**

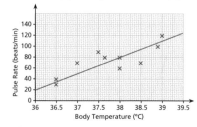

b) The estimate is not reliable as it is outside the data range. **[1]**

Pages 114–117 Practice Questions

Page 114

1. $\dfrac{500 \times 8}{40}$ **[1]**; $= 100 \pm 25$ **[1]**

2. 4kg carrots £1.96 **[1]**; 1kg = 49p **[1]**
3. £657 = (h × 35) – 218 **[1]**; $\dfrac{657 + 218}{35}$ **[1]**; = 25 hours **[1]**
4. a) $\sqrt{(6.2^2 - 3.6)}$ = 5.902541825, 2.6 × 0.15 = 0.39 **[1]**; 5.902541825 ÷ 0.39 = 15.13472263 **[1]**; 15.13 **[1]**
   b) 15.1 **[1]**
5. a) kilometres / km **[1]**
   b) grams / g **[1]**
   c) millilitres / ml **[1]**
6. 40km **[1]**
7. a) 9000 **[1]**
   b) 8800 **[1]**
   c) 8770 **[1]**

Page 115–116

1. $x = \pm 3$ **[1]**
2. a) Straight-line graph correct **[1]**; and curved graph correct **[1]**

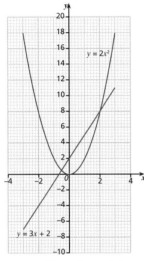

   b) $x = -0.5$ **[1]**; and $x = 2$ **[1]**

> When graphs of simultaneous equations are plotted, the x-coordinates of the points of intersection give you your solutions.

3. $2x + x - 2 = 1$ **[1]**; $3x = 3$ **[1]**; $x = 1$ **[1]**; $y = -1$ **[1]**
4. a) $(x + 4)(x - 2)$ **[1]**
   b) $x = -4$ **[1]**; $x = 2$ **[1]**
5. $3a + 2p = 220$ and $6a + 2p = 340$ (or equivalent working) **[1]**; $3a = 120$ **[1]**; $a = 40$ pence **[1]**; $p = 50$ pence **[1]**
6. Add equations to eliminate y **[1]**; $3x = 12$, $x = 4$ **[1]**; substitute value for x to find y **[1]**; $y = 1$ **[1]**
7. Multiply first equation by 2 and second equation by 3 **[1]**; $y = 2$ **[1]**; substitute value for y to find x **[1]**; $x = 3$ **[1]**
8. a) Accept values of x from 0.5 to 0.9 **[1]**; and –2.5 to –2.9 **[1]**

b)

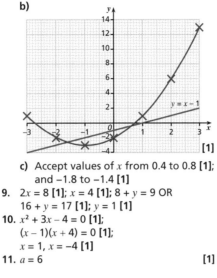

[1]

   c) Accept values of x from 0.4 to 0.8 **[1]**; and –1.8 to –1.4 **[1]**
9. $2x = 8$ **[1]**; $x = 4$ **[1]**; $8 + y = 9$ OR $16 + y = 17$ **[1]**; $y = 1$ **[1]**
10. $x^2 + 3x - 4 = 0$ **[1]**; $(x - 1)(x + 4) = 0$ **[1]**; $x = 1$, $x = -4$ **[1]**
11. $a = 6$ **[1]**

Page 117

1. a) 90° **[1]**
   b) 45° **[1]**
   c) 40° **[1]**
   d) 60° **[1]**
   e) 100° **[1]**
   f) 90° **[1]**

1. a) Triangle **[2]**
   b) (1, –2) **[1]**; (5, –2) **[1]**; (5, –4) **[1]**
   c) (–6, 5) **[1]**; (–2, 5) **[1]**; (–2, 3) **[1]**

Pages 118–121 Review Questions

Page 118

1. a) 805.975 ÷ 6.14 = 131.2662866 **[1]**; 131.27 **[1]**
   b) 131 **[1]**
2. a) Per day £84 (adults) + £35.70 (child) = £119.70 **[1]**; Holiday costs 9 × £119.70 **[1]**; = £1077.30 **[1]**

> They book 12 days but only pay for 9 because of the special offer.

   b) 5% of £1077.30 **[1]**; = £53.87 **[1]**
   c) 11.10am + 4 hours 20 minutes + 1 hour 34 minutes, flight arrives in Turkey at 5.04pm (17:04) in UK time **[1]**; Turkish arrival time = 7.04pm (19:04) **[1]**
   d) 3672 ÷ 8 = 459 **[1]**; 459 × 5 = 2295 miles **[1]**
3. a) 55g (+ or – 10 grams) **[1]**
   b) 2.5m (+ or – 1 metre) **[1]**
   c) 350ml (+ or – 50 millilitres) **[1]**

Page 119–120

1. $\frac{1}{2} \times x \times (x - 2) = 1.5$ **[1]**; $x^2 - 2x = 3$ **[1]**; $x^2 - 2x - 3 = 0$ **[1]**; $(x - 3)(x + 1)$, $x = 3$ **[1]**

> Remember, x is a length in this question so x must be positive.

2. a)

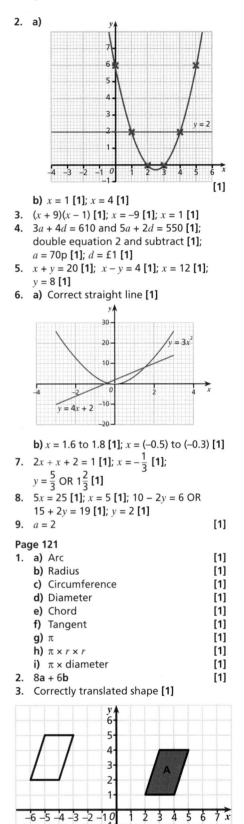

[1]

   b) $x = 1$ **[1]**; $x = 4$ **[1]**
3. $(x + 9)(x - 1)$ **[1]**; $x = -9$ **[1]**; $x = 1$ **[1]**
4. $3a + 4d = 610$ and $5a + 2d = 550$ **[1]**; double equation 2 and subtract **[1]**; $a = 70$p **[1]**; $d = £1$ **[1]**
5. $x + y = 20$ **[1]**; $x - y = 4$ **[1]**; $x = 12$ **[1]**; $y = 8$ **[1]**
6. a) Correct straight line **[1]**

   b) $x = 1.6$ to 1.8 **[1]**; $x = (-0.5)$ to (-0.3) **[1]**
7. $2x + x + 2 = 1$ **[1]**; $x = -\dfrac{1}{3}$ **[1]**; $y = \dfrac{5}{3}$ OR $1\dfrac{2}{3}$ **[1]**
8. $5x = 25$ **[1]**; $x = 5$ **[1]**; $10 - 2y = 6$ OR $15 + 2y = 19$ **[1]**; $y = 2$ **[1]**
9. $a = 2$ **[1]**

Page 121

1. a) Arc **[1]**
   b) Radius **[1]**
   c) Circumference **[1]**
   d) Diameter **[1]**
   e) Chord **[1]**
   f) Tangent **[1]**
   g) π **[1]**
   h) $\pi \times r \times r$ **[1]**
   i) $\pi \times$ diameter **[1]**
2. $8a + 6b$ **[1]**
3. Correctly translated shape **[1]**

Pages 122–131 Mixed Exam-Style Questions

1. Trapezium below line C shaded **[1]**
2. $3x^2(x + 2)$ **[1]**

Answers

3. $3x + 9 - 5 = 2x - 4$ **[1]**; $3x + 4$
$= 2x - 4$ **[1]**; $x = -8$ **[1]**

4. $\frac{5}{7} \times \left(\frac{1}{5} - \frac{2}{5}\right) + \frac{18}{7}$ **[1]**; $\frac{17}{7}$ **[1]**

5. $-5x^2 + 10x - 2y$ **[2]**

6. $4p^2y + 6py$ **[1]**

7. $x - 3 = -\frac{3}{5}$ **[1]**; $x = \frac{12}{5}$ **[1]**

8. $(x - 2)(x + 1)$ **[1]**

9. $30 = 60 - q^2$ **[1]**; $q = \sqrt{30} = 5.48$ **[1]**

10. a) $(x + 4)(x - 4)$ **[2]** (1 mark for each correct bracket)
b) $x = 4$ **[1]**; $x = -4$ **[1]**

11. Ethan is correct **[1]**; $r^2 = \frac{25}{\pi}$,
$r = \sqrt{\frac{25}{\pi}} = \frac{5}{\sqrt{\pi}}$ **[1]**

> You must square root both the numerator and denominator.

12. 6.8 **[1]**

13. a) $5 \times 3 \times 3$ **[1]**; $= 5 \times 3^2$ **[1]**
b) $3 \times 5 \times 7$ **[1]**
c) 3×5 **[1]**; $= 15$ **[1]**

14. $\frac{31}{6} - \frac{7}{3}$ **[1]**; $\frac{31}{6} - \frac{14}{6}$ **[1]**; $\frac{17}{6}$ **[1]**

15. $P = 1.1 \times 1.1$ **[1]**; $= 1.21$, 21% increase **[1]**

16. Dave's Dongles: £16.20 for six months **[1]**; £18 for following six months, total £205.20 **[1]**; Ian's Internet: £14.62 for four months **[1]**; £17.20 for following 8 months, total £196.08 **[1]**; Ian's Internet is cheaper **[1]**

17. a) 9.7×14520 **[1]**
b) 1408.44 **[1]**

18. $1.5 \times 10^8 \div 3.5 \times 10^4$ **[1]**;
4285.7 (to 1 d.p.) $= 4.2857 \times 10^3$ **[1]**

19. $S = \frac{\text{Distance}}{\text{Time}}$ **[1]**

20. 162° **[1]**

21. $3n + 5$ **[1]**

22. a) 2 **[1]**
b) $(10 \times 1) + (21 \times 2) + (3 \times 16) +$ $(4 \times 12) + (5 \times 6)$ **[1]**; $\frac{178}{65}$ **[1]**;
2.74 **[1]**

c)
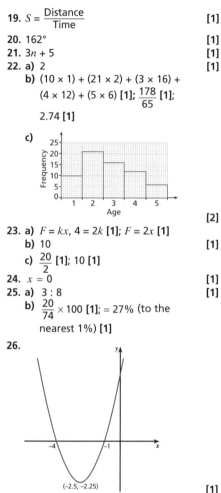
[2]

23. a) $F = kx$, $4 = 2k$ **[1]**; $F = 2x$ **[1]**
b) 10 **[1]**
c) $\frac{20}{2}$ **[1]**; 10 **[1]**

24. $x = 0$ **[1]**

25. a) 3 : 8 **[1]**
b) $\frac{20}{74} \times 100$ **[1]**; $= 27\%$ (to the nearest 1%) **[1]**

26.
[1]

27. $3x - 4 = 2x - 1$ **[1]**; $x = 3$ **[1]**; $y = 5$ **[1]**

28. a) 33° **[1]**; radius and tangent meet at 90° **[1]**
b) 57° **[1]**; triangle in semicircle or angle at circumference = 90° **[1]**

29. $\frac{1}{5}t^2$ **[2]**

30. $2 \times \pi \times 6 = 12\pi$ **[1]**; $12\pi \times \frac{115}{360}$ **[1]**;
$12\pi \times \frac{115}{360} + 12$ **[1]**; 24.0 (to 3 significant figures) **[1]**

31. $\frac{6}{11} \times \frac{5}{11}$ **[1]**; $\frac{30}{121} \times 2$ **[1]**; $\frac{60}{121}$ **[1]**

> The bead is replaced each time.

32. a) −2 and −1 **[1]**
b) The curve is symmetrical, so $x = -1.5$ **[1]**; $y = (-1.5)^2 + 3 \times (-1.5) + 2$ **[1]**; $y = -0.25$, $(-1.5, -0.25)$ **[1]**

Notes

Glossary and Index

Quadratic graph   the graph of a quadratic equation; produces a curved shape

Qualitative data   data that is non-numerical, e.g. colour of cars

Quantitative data   data that is numerical, e.g. number of cars

R

Radius / radii (pl.)   the length of a straight line from the centre of a circle to the circumference

Random   when each object / person has an equal chance of being selected

Range   the spread of data

Rate of change   the rate at which one quantity changes in relation to another; the change in the y-value divided by the change in the x-value

Ratio   the relative amounts of two or more things, shown in the form A : B

Rational number   a number that can be written exactly in fraction or decimal forms, e.g. $\frac{1}{4} = 0.25$

Reciprocal   the reciprocal of a number (n) is 1 divided by that number, i.e. $\frac{1}{n}$

Recurring   repeating

Reflection   a transformation that produces a mirror image of the original object

Reflex   an angle between 180° and 360°

Region   an area when dealing with graphical inequalities, often bounded by lines

Regular   shapes in which all the sides and angles are equal

Relative frequency = $\dfrac{\text{frequency of a particular outcome}}{\text{total number of trials}}$

Rhombus   a quadrilateral in which all sides are equal and opposite sides are parallel

Right-angled   a triangle with one 90° angle

Root   a root is the inverse operation to an index or power, e.g. $\sqrt{x}$ is the inverse of x^2 and $\sqrt[3]{x}$ is the inverse of x^3

Roots   the points of intersection, where the curve of a quadratic graph crosses the x-axis

Rotation   a transformation that turns an object; every point of the object is turned through the same angle about a given point

S

Sample   a proportion of a population

Sample space diagram   a probability diagram that contains all the possible outcomes of an experiment

Scalar   a numerical quantity

Scale factor   the ratio by which an object is made bigger or smaller

Scalene   a triangle with no equal sides or equal angles

Scatter graph   a statistical graph that compares two variables by plotting one against the other

Secondary data   data used for an investigation that has been collected by a third party

Sector   the area between an arc and two radii

Segment   the area between an arc and a chord

Sequence   a collection of terms that follow a rule or pattern

Significant figures   a means of approximating a number; the first significant figure is the first non-zero figure (working from left to right)

Similar   similar figures are identical in shape but differ in size

Simplify   to make simpler, usually by cancelling down

Simultaneous equations   two or more equations that are true at the same time and must therefore be solved together

Sine   a trigonometric ratio stating $\sin\theta = \dfrac{\text{opposite side}}{\text{hypotenuse}}$

Speed   a measure of how fast something is moving, i.e. $\dfrac{\text{distance}}{\text{time}}$

Sphere   a 3D shape that is round, like a ball; at every point, its surface is equidistant from its centre

Square numbers   the result of multiplying a number by itself, e.g. $4^2 = 4 \times 4 = 16$

Square root   the inverse of a square number, e.g. $4^2 = 16$ and $\sqrt{16} = 4$

Standard form   a number written in the form of $a \times 10^n$ where a is a number between 1 and 10; for large numbers n is positive and for numbers less than 1 n is negative

Subject (of a formula)   the subject of a formula is the variable that appears on its own; formulae can be rearranged to make different variables the subject

Subtraction   to take one value away from another; the inverse of addition; represented by –

T

Tally chart   a chart or table used to collect data, in which a mark is made for each item counted and marks are made in groups of five

Tangent   a trigonometric ratio stating $\tan\theta = \dfrac{\text{opposite side}}{\text{adjacent side}}$

Tangent (line)   a straight line touching the circumference of a circle in one place only

Term   in an expression, any of the quantities connected to each other by an addition or subtraction sign; in a sequence, any of the values connected to each other by a pattern or rule

Terminating decimal   is a finishing decimal, i.e. with a finite number of digits, e.g. 0.75, 0.36

Term-to-term rule   a rule that links a term in a sequence to the previous one

Theta (θ)   a Greek symbol used in trigonometry to represent an unknown angle

Transformation   an action that brings about a change to the position, size or orientation of a shape, i.e. translation, rotation, reflection and/or enlargement

Translation   a transformation that moves the object, but does not turn it

Trapezium   a quadrilateral with one pair of parallel sides

Tree diagram   a diagram that represents probability, with each branch showing a different possible outcome

Trigonometric ratios   the sine, cosine and tangent ratios used to calculate unknown angles and side lengths in right-angled triangles

Turning point   the point at which a curved graph changes direction, i.e. the highest or lowest point of the curve

Two-way table   a table that represents two variables in a set of data

V

Variable   a quantity that can have many values; often written as a letter

Vector   a quantity that has both magnitude (size) and direction (shown by an arrow)

Velocity   a compound measure that represents speed and direction of travel

Venn diagram   a diagram that uses circles to represent sets, with the position and overlap of the circles showing the relationships between the sets

Vertex / vertices (pl)   the corner of a shape

Vertically opposite   angles that are formed when two lines intersect; the four angles add up to 360° and two adjacent angles add up to 180°

Z

Zero term   the theoretical term that would come before the first term of a given sequence